NEVER
OUT OF
SEASON

Also by Rob Dunn

The Man Who Touched His Own Heart
The Wild Life of Our Bodies
Every Living Thing

NEVER OUT OF SEASON

How Having the Food We Want
When We Want It
Threatens Our Food Supply
and Our Future

ROB DUNN

LITTLE, BROWN AND COMPANY
NEW YORK   BOSTON   LONDON

Little, Brown and Company
Hachette Book Group
1290 Avenue of the Americas, New York, NY 10104
littlebrown.com

First Edition: March 2017

Little, Brown and Company is a division of Hachette Book Group, Inc.
The Little, Brown name and logo are trademarks of Hachette Book Group, Inc.

The publisher is not responsible for websites (or their content) that are not owned by the publisher.

The Hachette Speakers Bureau provides a wide range of authors for speaking events. To find out more, go to hachettespeakersbureau.com or call (866) 376-6591.

ISBN 978-0-316-26072-5
Library of Congress Control Number: 2016958939

10 9 8 7 6 5 4 3 2 1

LSC-C

Printed in the United States of America

To my family and those who feed us,
the farmers, the scientists, the keepers of the seeds

Contents

NEVER
OUT OF
SEASON

1

A Banana in Every Bowl

What a trifling difference must often determine which shall survive, and which perish!
> —Charles Darwin, in a letter to Asa Gray

O ur hunger has shaped the earth in much the way that the hunger of a caterpillar remakes a leaf.

Thirteen thousand years ago, each of our ancestors consumed hundreds of different kinds of plants and animals in a week.[1] Like wild chimpanzees and rats, they ate what could be found. Diets varied with the seasons. One berry in June, another in July. One insect when the rivers ran deep, other insects when they were dry. The species chosen also varied among cultures and regions. If you knew what a person was eating for dinner you could probably figure out what time of year it was and where that person was living. Not anymore.

With the spread of agriculture, the diversity of foods consumed globally was reduced. With the globalization of agriculture, it was reduced further and homogenized—made the same from one place to another. Humans now subsist on a declining diversity of foods. In 2016, the supply of calories to humans around the world was less diverse than it had ever been. Scientists have named and studied more than three hundred thousand living plant species. Yet 80 percent of the calories consumed by humans came from just twelve species and 90 percent from fifteen species. Our dependence on these foods has simplified the landscape of the earth. There are now more acres of corn than acres of wild grassland.

Global estimates of the composition of the average human meal hide the reality that the diets of people in some regions are even less diverse than average. In the Congo basin, for example, more than 80 percent of calories in people's diets come from a single crop, cassava (also known as yuca or manioc). In parts of China, rice accounts for nearly all calories consumed. In North America, more than half the carbon in the average child's body comes from corn—corn syrup, cornflakes, cornbread. Corn kids. And in the United States, the poorer and more urban those kids are, the more corn-dependent they are likely to be; that is, the more each and every one of their cells is likely to contain carbon atoms derived from either corn or sugarcane (another source of sugar).[2] Biodiversity provides the richness of life: a

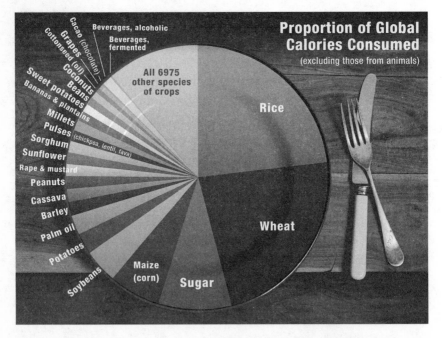

Figure 1. Proportion of global plant-based calories consumed by humans as a function of various plant sources. Sugar is derived from multiple plant sources, including sugar beets and sugarcane. The vast majority of domesticated plant species play a very small role on the global plate. *Data are drawn from Colin K. Khoury, et al., "Increasing Homogeneity in Global Food Supplies and the Implications for Food Security,"* Proceedings of the National Academy of Sciences *111, no. 11 (March 18, 2014): 4001–6.*

richness in species, ways of living, flavors, aromas, and attributes. What we now face is the opposite of biodiversity—a state of extreme monotony that is putting us at risk. Consider the banana.

.   .   .

On a plate, a single banana seems whimsical—yellow and sweet, contained in its own easy-to-open peel. It is a charming breakfast luxury as silly as it is delicious and ever-present. Yet when you eat a banana the flavor on your tongue has complex roots, equal parts sweetness and tragedy.

In 1950, most bananas were exported from Central America. Guatemala in particular was a key piece of a vast empire of banana plantations run by the American-owned United Fruit Company. United Fruit Company paid Guatemala's government modest sums in exchange for land. With the land, United Fruit planted bananas and then did as it pleased. It exercised absolute control not only over what workers did but also over how and where they lived. In addition, it controlled transportation, constructing, for example, the first railway in the country, one that was designed to be as useless as possible for the people of Guatemala and as useful as possible for transporting bananas. The company's profits were immense. In 1950, its revenues were twice the gross domestic product of the entire country of Guatemala. Yet while the United Fruit Company invested greatly in its ability to move bananas, little was invested in understanding the biology of bananas themselves.

United Fruit and the rest of the banana industry did what industries do. They figured out how to do one thing well—in this case, grow one variety of banana, the Gros Michel. Moreover, because it is difficult to get domesticated bananas to have sex (they are puritan in their proclivities, blessed with virtually no seeds), the Gros Michel was reproduced via suckers, clonally.[3] Cuttings from the best specimens were replanted. As a result, virtually all bananas grown in Guatemala, in Latin America in general, and around the world for export were genetically identical. Identical in the way that identical human twins are identical and even a tiny bit more so. For industry, this was great. Bananas were predictable. Each was like each other.

5

No banana was ever the wrong size, the wrong flavor, the wrong anything.

It is hard to overestimate how unusual the situation of bananas in the middle of the last century was—unusual not just in the history of humanity but also in the history of life. There is a patch of aspen trees in the Wasatch Mountains of Utah that many argue is the largest living organism on earth. It comprises some thirty-seven thousand trees, each of which is genetically the same as the other, and the argument goes that the trees, collectively, represent a single organism because they are identical and connected by their roots. But requiring pieces of an organism to be connected in order to be considered part of a collective is arbitrary. The ants in an ant colony, for example, are clearly part of the colony, even when they're not physically in the nest. All this is to say that an argument can be made that large groups of genetically identical plants, even if not connected, may reasonably be considered a single organism. If one makes such an argument, the banana plantations of Central America in the 1950s were not only the largest collective organism alive at that point, they also may well have been the largest collective organism ever to live.

Economically, growing just a single clone of bananas was genius. Biologically, it posed problems. These problems had already been noted, for example, in the British production and export of coffee in the 1800s. At that time, the British drank coffee, not tea. They drank coffee exported from their colony Ceylon (now Sri Lanka). Early on in Ceylon, coffee plantations were planted among wild forests.[4] When the British took Ceylon from the Dutch in 1797, they began to expand coffee production on the island. Investment in the coffee plantations by the English, both at home and abroad, "was unlimited; and in its profusion was equaled...only by the ignorance and inexperience of those to whom it was entrusted." As the demand for coffee increased, it was planted in large monocultures—that is, vast areas of only a single variety of tree. Coffee on one hill, coffee on the next. Not a taller, wilder tree to be seen. There were 160,000 hectares of the central uplands planted in coffee. The coffee brought real affluence—banks, roads, hotels, and luxury. It was an unbridled success, or seemed to be.

Harry Marshall Ward, a British fungal biologist visiting Ceylon in 1887, warned farmers that farming such large plantations of a single variety of coffee would cause problems. Pests and pathogens, once they arrived in the plantations, would devour them. This was, he thought, particularly true of coffee rust, which was already present in Ceylon, but it would also be true of any other pest or pathogen that arrived. Nothing would stop such an organism from quickly devouring all the trees, since they were all of the same variety—and thus equally susceptible to whatever threat might arise or arrive—and planted very close together. This is exactly what happened. Coffee rust wiped out the coffee of Ceylon and, subsequently, much of the rest of the coffee of Asia and Africa.[5] Coffee growers replanted with tea.

Ward had predicted that the coffee of Ceylon would be devastated. As the plantations of bananas expanded across the American tropics, scientists made similar predictions. These scientists noted that in the native range of bananas lived a great diversity. There were big ones, small ones, sweet ones, sour ones, hard ones, soft ones, bananas as dessert, and bananas—plantains, really—consumed as sustenance. In those same regions one could also find an extraordinary diversity of pathogens. But in the cultivated world of bananas, the scientists pointed out, because a single genetically identical variety of banana was planted everywhere, were any banana-attacking pathogen to arrive, it would mean trouble. Any pathogen that could attack a single banana plant, even one, would be able to kill all of them. If the banana companies had listened to these warnings, they might have planted a diversity of banana varieties or a variety that would be resistant to the most likely pathogens. But why would they? The single clone of the Gros Michel banana was the most productive anyone had ever found. Planting anything else would mean losing money.

Then the inevitable happened. A malady arrived—Panama disease (now more often called fusarium wilt), caused by the pathogen

Fusarium oxysporum f. sp. *cubense*. Panama disease started to wipe out banana plantations in 1890.[6] Nothing precluded its spread or even promised to slow it. Seen from above, the plantations across Latin America started to look like the lights had been turned off. Patches of bright green went black. Whole landscapes went black. In the Ulua valley of Honduras alone, thirty thousand acres were infected and abandoned within the first year in which Panama disease arrived. Nearly all the banana plantations in Guatemala were devastated and, once devastated, abandoned, because it was quickly figured out that the pathogen, having arrived, could lurk in the soil for years (or even, as we now know, decades).

United Fruit Company's leaders believed that if they were able to find another banana, one that vaguely resembled the Gros Michel but was resistant to the pathogen, it could be planted on the abandoned land and the banana empire could be restored. This plan, however, was based on a farcical set of assumptions. It assumed that consumers would simply accept whatever banana you sold them as long as it looked more or less the same. In addition, it overlooked the reality that no replacement banana had yet turned up—no good option, anyway. The only banana that seemed both pathogen-resistant and similar to the Gros Michel was a banana called the Cavendish. The Cavendish tasted very different from the Gros Michel. It had "off flavors" and was less sweet. What it had going for it, though, was that you could plant it even where Panama disease was present in the soil and it wouldn't die (and it still doesn't).

Over the next several years, the Cavendish banana would prove to be the only banana that both looked like the Gros Michel and would resist Panama disease. So it was that without any other real options, and having helped to overthrow a democratically elected government[7] so as to continue to be able to produce cheap bananas, the United Fruit Company started to plant the Cavendish across hundreds of thousands and then millions of acres. They then began to export it to the United States, along with a massive advertising campaign lauding the benefits of the banana. It worked. Just as the British had earlier switched from coffee to tea (substituting one caffeinated drink in a cup for another), Americans switched from the

Gros Michel banana to the Cavendish. The advertising was so good that the new banana, the Cavendish, was even more successful commercially than had been its predecessor, the Gros Michel. Bolstering the Cavendish's sales was the shift of American populations to cities, where the connection between what consumers bought and what grew well locally had been severed. Sales of the Cavendish banana were strong, and they continue to be. It is with very few exceptions the only kind of banana you find in stores outside the regions where bananas grow.[8] Its success fuels the economies of whole countries. It is the biggest export of Costa Rica, Ecuador, Panama, and Belize and the second most valuable export for Colombia, Guatemala, and Honduras.[9] If you were born after 1950, you are unlikely to have ever purchased any banana other than the Cavendish clone—other than what is now the world's largest organism. To the extent that anyone worried about diseases affecting the Cavendish, it was because of black leaf streak (*Mycosphaerella fijiensis*), which was not nearly as bad as Panama disease. Panama disease, meanwhile, had become a thing of the past. The Cavendish remained resistant in part because the pathogen itself is not very diverse and so relatively unable to adapt.

. . .

Industry, we learn from the story of the Cavendish banana, will plant the crop that grows most easily and supply it to us whenever we want. It will encourage us to want it all the time. It will tend to plant crops in ways that produce the greatest yield, even if that mode of production has costs; even if it also puts the very crop the industry depends on at risk. Cavendish bananas are all genetically identical. Each banana you buy in the store is the clone of the one next to it. Every banana plant being grown for export is really part of the same plant, a collective organism larger than any other on earth, far bigger than the clonal groves of aspens. This giant organism is now at risk of exactly the same sort of population crash that befell the Gros Michel, and a new strain of *Fusarium,* a close relative of the pathogen that causes Panama disease, has evolved. It can kill both Gros Michel and Cavendish bananas. This strain has already spread from Asia to East Africa and seems likely to make its way to Central America. This should be extremely worrisome. But what should be more worrisome is that the same is true of most of our crops, most of the plants that we most depend on, a list of species that is shockingly and increasingly short.

The simplification of the agricultural world and our diets has come with benefits. They are the same benefits that accrued to the United Fruit Company (rebranded in 1984 as Chiquita Brands International, a.k.a. Chiquita)—the ability to produce a large amount of food on a given area of land. In concert with the homogenization of agriculture, we have figured out how to grow more food per acre than ever before—ten times more food than ten thousand years ago, perhaps a hundred times more than fifteen thousand years ago. As a result, a smaller number of people on earth go hungry today than at any other moment in the last thousand years. Modern science has brought us food in abundance, just as it brought the United Fruit Company affluence. Yet this abundance, like the affluence of modern banana companies, is tenuous, dependent on our ability to protect the very few species on which we now depend. The problem is that nearly all those key species are in trouble, because in simplifying the production of our food we achieved short-term benefits at the expense of long-term benefits—and, for that matter, at the expense of long-term sustainability.

The problem we face is the consequence of the preferences of our brains, reinforced by the incentives of industry. We live in a thoroughly modern world with brains and bodies that evolved in an environment where sweets, fats, proteins, and salt were all hard to get. We have simple ape brains and simpler ape nervous systems. Our ancestors evolved taste buds that rewarded them when they found food that provided these necessities. Our environment has changed. Our needs have changed. But our taste buds remain the same. We experience pleasure when we eat these substances, our body's way to reward us for having found them. Our brains, meanwhile, are wired to spot shiny, bright fruits. As a result, the world we were most likely to create is one in which our foods appeal simply to these ancient preferences. This is precisely what we have done and precisely what one encounters in the grocery store, where the foods in the greatest abundance are now perfectly matched to our ancient needs despite our modern waistlines. Inasmuch as we demand (or at least buy) the same things regardless of the time of year, the foods in the grocery store are never out of season. What's more, whereas the fruit and vegetable aisles of some grocery stores are relatively diverse, the vast majority of the calories in our diets come from the processed foods found in the rest of the store, foods that can stay on the shelf long beyond the seasons of the plants (or animals) from which they are made.

Globally, we favor the crops that best satisfy our ancient needs at the lowest cost, regardless of how far they might have to travel and regardless of the season. The more urban our civilization becomes, the more disconnected it becomes from the life on which we depend and thus the more extreme our demand for simple products regardless of the season. The crops that are expanding—in terms of the area over which we plant them—are not those that are the most flavorful or nutritious but rather those that are used to produce sugar (sugarcane, sugar beets, corn) and oil (oil palms, olives, canola).[10]

That we have created such a simple world seems dissatisfying, but just because something is dissatisfying doesn't mean it won't suffice. Theoretically, we could live off of a diminishing number of crops. We could even get by on a single crop. Potatoes, for example, provide nearly all the nutrients we need, as do cassava and sweet potatoes. But just as our demand for a few basic foods whenever we want them was predictable, so, too, were the problems these crops are now facing. The more we feed ourselves according to our most primitive desires, the more we create a world dominated by just a few productive crops—crops that are threatened by their very commonness. Even coffee is at risk again. Having learned nothing from Sri Lanka, we have once more planted varieties of coffee that are susceptible to coffee rust in large plantations, and the rust is back. That these crops are nearly all at risk today from pests, pathogens, and climate change is not a fluke. Given our preferences, it was nearly inevitable.

The risk to our crops comes in direct proportion to the ways in which we have simplified agriculture. Nearly every crop in the world has undergone a very similar history—domesticated in one region, then moved to another region, where it could escape its pests and pathogens. But these pests and pathogens, in our global world of airplane flights and boat trips, are catching up. Once they do catch up, there are only very few ways to save our crops, and all of them depend on biodiversity, whether in the wild or among traditional crop varieties. This was true with the banana. Saving banana production around the world depended on finding the Cavendish banana, which relied on the work of the farmers that produced and grew it in the first place. Saving the banana when the Cavendish collapses will depend on our finding yet another variety and having similar luck. Alternatively, someone might be able to breed a new, resistant banana using some mix of new technologies and ancient varieties. But if they are going to do so, it will need to be soon.[11]

The more we heed our basic instincts for cheap sugar, salt, fat, and protein in whatever form we want it, whatever time of year we want it, the more we create a simple agricultural world and the more we will depend on the diversity of life with which that same agriculture competes on a finite planet. This book is the story of scientists racing to save the diversity of life in order to save our crops and in order to save *us*. It is the story of a puzzle we must solve. The ancient rules of life leave us relatively few ways to arrange the pieces.

2

An Island Like Ours

The sauce of the poor man is a little potato with a big one.
—Anonymous

Men sprinkled holy water on their potatoes; they buried them
with religious medallions and pictures of Christ and the Virgin
Mother. Nothing worked. God had turned away.
—John Kelly, *The Graves Are Walking*

In July of 1846, after a long winter, the fields of Ireland were as
green as a golf course and covered with the shoots of potato plants.
Then, in forty-eight hours, everything changed. From one end of Ireland to the other, the potatoes died. Near Cork, a traveler found a
solitary man in a field, singing. When asked what he was doing, the
man said that all his potato plants were dead, blackened and oozing.
His livelihood was gone, as were his options. What else could he do?
Near him a woman scraped the ground of another field, her body
bent, her hands clawing. Beside the woman were a few tiny, oozing
potatoes. She planned to cook them for her children. She had nothing else. No wheat. No carrots. The cow had been sold. The same
thing was happening across nearly all of Ireland to millions of desperate Irish at the beginning of what was about to become one of the
worst tragedies in modern human history.

The scale of the horrors of the Irish potato famine is almost
beyond our ability to conceive. The young died first, then the old,
then everyone else. People died in the ditches where they slept for the

night, en route to what they hoped might be someplace better. They died in their fields. Whole villages disappeared. More than a million people would die before it was all over—a million in Ireland, that is. Others left Ireland on ships, only to face, nearly as often as not, death themselves. The magnitude is numbing. But what is perhaps most astonishing about the famine as it relates to our lives is that more crops are at risk of devastation today because of pathogens and pests than were at risk when the potato famine occurred. The potato famine was not the last ancient plague but rather the first truly modern one. And whereas the threat from the potato famine was regional, the threat we now face, in our far more connected economy, is global.

The potato famine was caused by a disease we now call late blight (and that was then called potato murrain).[1] Late blight was first noted in New York in 1843. Where it arrived, potato plants died. It spread to Pennsylvania within the year and left, in its wake, even more dead plants. From the perspective of farmers in Pennsylvania and New York, the late blight fell from the sky. It rained down like a curse. The next spring, potatoes were dying as far north as Vermont. In the spring of 1845, the late blight was in Newfoundland, Canada. Then, later that year, it was in Belgium. Once the blight was in Belgium, its rate of spread increased, its waypoints measured in months rather than years and then weeks rather than months. The late blight was in France by July. By August it had reached England.

In the United States, potatoes were a relatively small portion of the average diet, and so while the losses were great to individual farmers, the collective loss was modest. In Europe, particularly northern Europe, things were different. Between 10 and 20 percent of people in the Netherlands, Belgium, Poland, and Prussia ate little solid food other than potatoes. The arrival of the late blight in these regions threatened the sustenance of many families. The death of potatoes in places the late blight had arrived was so extensive that newspapers could discuss little else. Flanders lost 92 percent of its 1845 potato crop, Belgium 87 percent. The Netherlands lost 70 percent. Even in these countries, each of them far less dependent on the potato than was Ireland, the consequences were dire. In the Netherlands the relatively well-to-do were said to "live on the herbs of the field" in the fall of 1845. This was still in the fall, before the long winter. Famine lurked in the small towns and homes across rural northern Europe. The real worry, though, was the small but densely populated island of Ireland.

In 1845 the Irish were more dependent on the potato for sustenance than any other group of people in Europe and, for that matter, any other people on earth, even Andeans.

This dependence of the Irish on potatoes was new and partially the result of chance—i.e., the fact that the potato arrived in Ireland from the Americas (where it was native) in the first place. The dependence on potatoes was also partially attributable to the challenges of farming on a cool, wet island where few crops other than the potato grow well. But perhaps the biggest reason that the potato came to dominate was the system of land ownership. In Ireland in the nineteenth century, Protestant barons of British descent owned enormous estates on which middlemen rented land to the masses. The masses paid rent in part by giving their landlords their agricultural surpluses, though "surpluses" is a misnomer. It would be better to say that the tenants gave their landlords a fixed amount of what they grew, which was then sold to the growing urban population in Dublin, Belk, and Cork as well as to urban populations in England, then the tenants themselves consumed the surplus. Given this land system, success for the average Irish family was measured in terms of producing enough food to survive after the landlord took his share. The crop that produced the most food per acre was the potato.

With each generation, Irish dependence on the potato increased; the Irish were locked in a cycle. The potato and, more specifically, the lumper potato, provided complete nutrition, particularly when combined with milk—complete nutrition that prior to the arrival of the potato in Ireland was lacking. Infant mortality decreased with the farming of the potato. Life expectancies increased. The Irish population boomed, as did populations in other parts of Europe where the potato had become the dominant crop. But as populations boomed, land had to be further subdivided, and as a result families became even more dependent on the potato, the only crop that could sustain them on ever-smaller pieces of land. By the early 1800s, a poor tenant family was likely to have little more than an acre. The only crop that provided for a family on so little space was the potato, and no one would dare plant much else because it would mean having too little to eat. The Irish were trapped eating potatoes, and they ate a lot of them. On that typical acre, by 1845, the average adult in western Ireland may have been consuming as many as fifty to eighty potatoes per day.[2] They often did not have clothes or shoes. They lived in houses carved out of sod. They were penniless and yet, thanks to the potato, nourished.[3] This was the luck of the Irish in the early 1800s.

In looking back at the Irish in 1845 it is easy to think of them as backward. But they were the opposite: a culture sustained by the newest approach to agriculture, one in which a single variety of a single crop is planted on a large scale, fertilized, and consumed disproportionately. The Irish represented, in their dependence, a potential version of our future. As of early 1845, it was still a hopeful future. The late blight, whatever its cause, had not made it the eighty miles across the Irish Sea from England,[4] and so the lumper grew in the dark soils of thousands of fields, as rich and sustaining as it ever had been.

· · ·

Meanwhile, the scientists best able to solve the problem of the late blight of potatoes were arguing. No one could agree as to just what caused the disease, and so a contest was held to find the best essay that explained its cause.

Many hundreds of entries were submitted. From among the essays, first, second, and third prize winners were selected. All three winning essays argued that the cold, wet spring weather was to blame, though most agreed it could well be some combination of weather and bad seeds. Of course, there was some modest disagreement about just what aspect of the weather was responsible. Was it the rain or the cold rain? Or maybe it was the cold rain followed by more cold. When pathogens such as fungi were mentioned, they seemed often to go hand in hand not with these two hypotheses but instead with an odd list of wild explanations offered by those who failed to know better. The dust from sulfur matches. Pollution. Volcanoes. Airborne miasma from outer space. Or fungus, though to many the idea that a fungus could be responsible for such a sweeping tragedy was plain silly, certainly not the topic of a winning essay.

Yet the advocates of the fungus hypothesis included a handful of scholars from across Europe. The French mycologist Abbé Edouard Van den Hecke, vicar-general to the bishop of Versailles, suggested that a fungus was to blame in the newspaper *L'Organe des Flandres* on July 31, 1845. Observing the organism under a microscope, he marveled at its ability to disperse itself. If this organism was causing the blight, Van den Hecke noted, it was vital to remove infected plants lest the organism and disease spread. This was, he went on to note, the importance of determining the role of climate, bad seeds, and pathogens. If the cause was climate, one cursed the gods and waited. If it were bad seeds, one got other seeds. If it was a pathogen, well, one had to figure out how to stop its spread, keep it from growing, and, once it had grown, kill it.

Then on August 14, Professor Martin Martens, of the Catholic University of Louvain, chimed in. Martens thought that the cause was indeed the same organism studied by Van den Hecke. He, too, noted that it was vital to remove diseased plants to prevent the spread of the offending organism—the pathogen, we would now say. He described the biology of the creature in more detail. It attacks, he said, the leaves, "especially on their lower faces." On August 19, in the *Journal de Liège,* Marie-Anne Libert, a self-taught but respected mycologist, named the organism and described its details. She wrote about its appearance, biology, and life history. She considered it to be a form of *Botrytis farinacea,* which she suggested should be named *Botrytis devastatrix,* where "devastatrix"[5] was intended to reflect the organism's consequences.

The drumming of minor proclamations continued. In Belgium, another respected scientist also concluded that the cause was a fungus. On August 20, Charles Morren published an article in which he systematically ruled out all potential causes of the disease except the organism *Botrytis devastatrix*. He then claimed that fungi and related organisms also cause ergot of rye, wheat rust, oat smut, and corn smut. Morren, too, offered clear suggestions for getting rid of the problem. Remove and burn diseased plant material. Get rid of any infected potatoes. Dip the superficially healthy potatoes in copper sulfate,[6] lime, and water. Plant in the fall rather than in the spring or summer.

Then in the summer of 1845 the late blight arrived at the doorstep of one Miles J. Berkeley, officially the vicar of King's Cliffe, in Northamptonshire, but unofficially among the most accomplished scientists in England. He'd made discovery after discovery about the species living around his home, discoveries others had missed. Berkeley studied many species — algae, plants, even some animals — but fungi were his love. He spent thousands of hours gathering mushrooms.[7] During his life he collected more than ten thousand species,[8] at least five thousand of which were new to science and so named by him. When his interest outlasted the sun's light he worked at night or in the early morning by candlelight. He was very curious as to whether a fungus might be destroying his potato plants, but until he had seen the diseased plants himself he had his doubts. Berkeley collected a few diseased potatoes from fields near his house, brought them into his workroom, and started to look for fungi. If anyone in the world could identify what was causing the blight with certainty, it was Berkeley. Scientists of his time regarded him as a great expert in the biology of fungi and crops. Scientists of our time regard him as the father of plant pathology — or at least as its favorite uncle.

Yet in considering the hypothesis that fungi were to blame, Berkeley was not just figuring out the potato's late blight, he was also offering one of the first efforts to establish germ theory. We take germ theory, the idea that an illness of an animal or plant could be caused by a smaller organism, for granted. We wash our hands. We cover our mouths. Such precautions seem obvious to us, but they were not obvious in 1845, nor had they been at any previous point in the long history of humans.[9] At the time of the potato famine, germ theory would not be applied to human (or crop, as it would turn out) disease for decades.[10] But as Berkeley examined the blighted potatoes in front of him closely, he agreed with others who had identified the cause as a minute organism of the genus *Botrytis*.[11] He observed that the late blight did not arrive until this *Botrytis* had. The cause, he was nearly sure, was *Botrytis infestans*—another name for *Botrytis farinacea* and *Botrytis devastatrix*—where *infestans* refers to the ability of the creature to infest. The good news, the very good news, was that with the culprit identified, a course of action could be identified as well. Berkeley started to work through what that course might be. As an esteemed man of science, he could add his voice to the chorus of those who thought fungus was to blame, and people could get down to the business of treating plants and preventing them from being infected in the first place.

Unfortunately, things were not so simple. For one thing, there was still strong societal pressure to blame the disease on the weather or bad seeds. No one outside the afflicted regions would want to buy the potatoes if they possessed some organism capable of killing other potatoes. The ancient fear of the other now had a new face. In addition, the idea that a fungus caused disease was just too radical to consider at a time when science was conservative (as it for the most part remains). Both economic politics and the status quo were against the fungal hypothesis, so much so that Camille Montagne, a former surgeon in Napoleon's army and one of the scientists who initially argued that a fungus was to blame (it was Montagne who first used the species name *B. infestans*), backed down.[12] He was worried that he would not be elected into the Academy of Sciences in France if he continued to advocate for the fungal cause. The lives of millions of Europeans were at stake, but the Academy of Sciences was the Academy of Sciences. The pressure was such that even as the evidence in favor of the pathogen hypothesis increased, the support for it began to disappear. By late summer of 1845, although the late blight continued to spread and a number of individuals had advocated the idea that fungi were to blame, Morren and Berkeley were the only internationally well-known scientists actively advocating the pathogen hypothesis. Morren was the more vocal of the two but less highly regarded. Berkeley was very highly regarded, but is sometimes described as having been so reserved as to be nearly mute.

. . .

Then it happened. On September 6, 1845, two newspapers announced the arrival of the late blight in Ireland—perhaps from England, though who could say for sure? On September 13, just seven days later, *The Gardeners' Chronicle*[13] reported that in parts of Ireland the fields were already completely destroyed. The disease was spreading nearly as fast as a man could skip. By October no field in Ireland was beyond the late blight's reach. Within two months, more than three-quarters of a million acres of potatoes were simply gone, turned to stinking, black rot.

Those who walked these fields described first and foremost the smell. One got used to the odor of cow dung or even the acrid sting of chicken poop in your nostrils, but not this. Everywhere potatoes had been planted, their infected tubers and stems gave off a sulfurous stink; it was, some said, as though hell were leaking out of the ground.

By January, famine was "striding nearer every day like a wolf in search of prey," as a poem in the Irish newspaper *The Nation* put it. Remnants of the previous year's crops remained at some homes, but they were diminishing. By March farmers were raiding one another's fields, searching for missed roots or whatever else might be had. The scenes were postapocalyptic: "Men, sucking the blood from the neck of a living cow, seaweed on the boil; grass-stained mouths and hands; women running an anxious hand over a sleeping child to see if she still breathed."[14] Peasants who had never had much sold what was left—their clothes. It was not uncommon to see children and families, nearly naked in the cool March air, walking out to work what was left of their fields.

By the spring of 1846, the nightmare seemed to be drawing to a close, and most families thought they would soon have enough to eat. In June the grains and vegetables were coming in strong, and the potatoes were in the ground. The blight of 1845, whatever its cause, had been an anomaly. The Irish farmers just had to make it until the final harvest in October, the harvest on which families would survive through winter. Then in July, a bad omen fell from the sky. The torrential rains that had been associated with the late blight the year before, here and there across Ireland and then everywhere, were starting again.

By the summer of 1846 in Ireland, apprehension had turned to horror. The rains continued, and as they did the late blight spread to even more fields than it had the year before. The potato crop was lost. The already hungry began to starve. During the winter, fever became common. Typhus. Relapsing fever. Thousands began to die.

In 1847 the story repeated itself again. The rains came. The crop disappeared. The hungry who managed to survive had been hungry three years in a row. One William Wilde described absentmindedly staring at the lovely fields around him in a valley one night and, two nights later, after two days of rain, looking out at a black and stinking horror. An entire valley dead. Nor was this landscape unique: in just a few days most of the crop of Ireland was destroyed.[15]

Ireland stank. In 1845 it was just the odor of the dying potatoes. But by 1847 this odor was accompanied everywhere by that of human bodies, the naked and starving families alongside the road. Others lay half dead, half alive, prostrate in a state of mute yearning. Everywhere human bodies gave off a sweet odor of living decay. Then there was the more overpowering smell of the dead. In fields, huddled together on beds in homes, piled in the mass graves where they were lowered, one after the other, day after day. New kinds of coffins were invented—those with drop-away bottoms that could be used again and again as bodies were dropped one on top of another to save space and to avoid the need to build more coffins and dig more holes.

By August of 1847, those with guns had hunted out the last of the animal life that remained. By September, the guns and any remaining bullets were being used to hold up landowners and steal from them. By October, even the bullets were gone. By November, thousands had died. By December, tens of thousands, then hundreds of thousands. The British government did not help, nor did the government administrators in Dublin. Most landowners did not help, either, but they did pay taxes based on the number of people living on their land. It was these taxes—not the death and devastation consuming the country—that motivated the landowners to take real action. They began to tear down the houses of their tenants who could not pay rent in the hope that the hungry would move away and reduce the tax burden on the land. When the tenants did not, many of the landowners coerced the poor into boarding ships for the Americas. Bodies, half alive, were stacked on top of each other in the holds of ships. It is an image that now evokes the ships of immigrants from Africa and Syria heading to Europe or Cubans on rafts floating toward Florida. More than a million Irish are estimated to have fled Ireland during or after the famine. Many of those people—men, women, and children—died en route or in the weeks or months after arrival. And still there was no solution. The blight could, it seemed, go on forever, until nothing at all was left in Ireland.

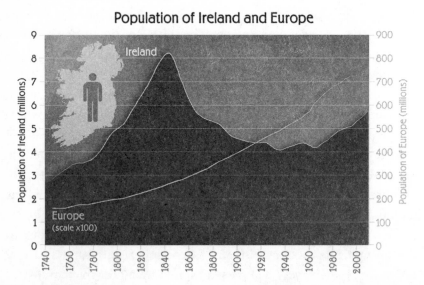

Figure 2. Population trends in Ireland and Europe as a whole. In the mid-1700s and late 1800s, potatoes contributed to the rapid rise in population in northern Europe, especially Ireland. But with the arrival of late blight, potato populations crashed, triggering the emigration and death of much of the Irish population. *Numbers are drawn from census data.*

. . .

Perhaps the hardest thing to conceive of is that the demon that stalked the Irish fields could afflict us today. Like that of the nineteenth-century Irish, our diets have become ever simpler and more dependent on a small number of species. Today, just as in the 1800s in Ireland, this dependence results in part from our need and in part from our choices. We need ever more food from each acre and so are bound to those crops that produce the most. Just as it was for the Irish, each time a child is born our reliance on our most productive crops increases. Corn in North America. Wheat in Europe. Cassava in Africa. Rice in Asia.

We are like the Irish in another way, too. The productivity of their fields came to depend on both irrigation and fertilizer, which they used to farm hills, to farm bogs, to farm everywhere. The Irish potato farmers were farming the first really modern crop. Our similarity to the Irish means that we must learn every lesson we can from their tragedy. We must learn especially from the way this tragedy came to an end.

In 1847, in Ireland, even after half a million people died and another half a million fled, the task of dealing with the late blight remained. No solution had yet been implemented. Nor had any other crop taken over in the absence of the potato. So long as the Irish and Europe in general blamed only the weather and fate alone, all there was to do was wait.

The debate about the cause of the late blight had essentially stopped. The group who thought it was a fungus went quiet. Few new articles appeared. Even Berkeley, who so clearly saw the fungus on the leaves near his farm, began to write with less certainty, giving voice both to the fungal theory and to other ideas. Many came to the mistaken belief that Berkeley had repudiated his own fungal hypothesis.[16] As for those who felt more strongly, if they existed in those years at all, they were quiet, which is all the more horrific, because we now know they were right.

The late blight that struck Europe beginning in 1845 is an oomycete.[17] Oomycetes were at the time still considered to be fungi (their common name, water mold, reflects this history). But oomycetes are not fungi, nor are they, as they are sometimes described, algae. They are an ancient life form you may never have heard of. Yet we now know that a larger number of oomycetes harm crops—and, by virtue of our dependence on crops, humans—than do any other organisms. One species of oomycete causes sudden oak death. Another causes the root rot disease of soybeans. Then there are the pepper murrain and the downy mildew of grapevines.[18] The oomycete that causes late blight is the same beast that Berkeley, Morren, and others called a minute fungus. Nearly everything they said about its biology was true. The spores of the organism travel through the air and land on plant leaves. Once there, they send snakelike tubes into the cells of the leaves. The oomycetes then feed on the insides of the leaves. The leaves begin to blacken and stink as the oomycetes turns them to goo and, in larger quantities, more oomycetes (which are visible as a white fringe on the blackening leaves).[19] From inside the plant, the oomycetes send more spores out through the stomata, or breathing holes, of the potato plant. These spores colonize more plants, and the process continues anew, as it did from plant to plant across continental Europe and from continental Europe to England and on to Ireland. The spores spread better when it is windy, and the oomycete grows faster when it is wet. Both these things would have led the oomycete to do better in 1845, 1846, and 1847 than it did in other years. But the spread was almost certainly helped by something else—farmers dragging plants from place to place. All these details of the biology of the organism had been noted by 1846. They had been noted in great detail; they were just being ignored. For reasons we can go on to debate, Berkeley and those who agreed with him failed to communicate what they *knew* to be true in ways that would convince others.

The historian Steven Turner tells us that "the fungal hypothesis had been largely discredited by the end of 1847." As a result, it was also ignored in 1848, 1849, 1850, 1851, and for the following thirty years, during which no solutions emerged to late blight. The tragedy continued for decades. The potato became rarer and harder to grow. Slowly, other crops were planted in greater quantities. Also, each year until the 1880s, there were fewer mouths in Ireland to feed. Across these generations of tragedy, scientists, or at least some scientists, knew the cause of the blight and even thought they knew how to control it. The loss of lives during the blight was almost inarguably one of the biggest failures in of communication, at least in the field of science, across the history of humanity.

What it would take to convince people was not more data: it was time and an experiment that unambiguously demonstrated and communicated what was by then, for all practical purposes, already known. As we all learned in grade school, Louis Pasteur performed experiments demonstrating the ability of pathogens to cause human diseases. In doing so, he offered a way to prevent and treat many terrible and deadly diseases. These experiments ushered in the modern study of human disease and offered a framework for establishing what was and was not the causal agent of any particular malady. But it was not Pasteur who did this work first. The very first such work was done by the German scientist Heinrich Anton de Bary, and it was done to understand the late blight oomycete.

De Bary was a "small, nervously-eager young professor of botany."[20] In 1853 he wrote a book on the smuts and rusts of wheat, oats, and rye. At the time of the book's publication, de Bary was just twenty-two. He then spent the next years studying the sex lives of algae and, to a lesser extent, fungi such as the powdery mildew (*Erysiphe cichoracearum*) of dandelions. During those years, he might have considered the potato blight. He did not. It was not until 1860 that de Bary took the organism that Berkeley hypothesized was the root cause of the potato famine and inoculated potatoes with it. This approach is now so much a part of our modern understanding of the world that it seems obvious. It wasn't then. De Bary was the first to try to test whether a pathogen caused a disease through experimental inoculation. The potatoes, even though it was a good, warm, dry year, died. If a "eureka" moment existed in the story of the potato blight, this was it. De Bary had proved that the late blight was caused by the late blight organism. By extension, de Bary's result meant that Berkeley was right, but the discovery was made at the least satisfying possible moment, only after so many Irish had died or fled. Still, once de Bary proved the blight was to blame, everything made sense. The potato late blight was worse in wet years than dry years. It was worse when the plants were inbred than when they were not. But it was caused by an oomycete, a funguslike organism, an organism that in theory could be killed.

Once de Bary confirmed the cause of the late blight, the only thing left was to get rid of it. The solution that appeared was the use of pesticide. Irish potatoes were modern before the famine in the sense that with fertilization, they could be planted in places beyond where they would naturally grow. They were modern after the famine in another sense, in that they could be grown only with the use of pesticides.

One of the pesticides used today to kill late blight on potatoes is a mixture of copper sulfate and lime. In a horrible twist, it is the same pesticide that was advocated multiple times by supporters of the fungus hypothesis as early as 1843. Morren himself wrote in 1845 that copper sulfate could be used to destroy the blight. It could not, he said, "be too strongly recommended."[21]

Had Morren's advice been followed, hundreds of thousands of human lives might have been saved, perhaps millions. It was not heeded. What's worse, the other individuals who, apparently independently, also voiced the value of copper sulfate, a New York judge with the last name of Cheever (whose first name has gone unrecorded) and James E. Teschemacher of Boston, were also ignored. Instead it was not until 1883 that those who advocated the use of copper sulfate were listened to, and even then it was not a scientist who had the initial insight but rather a farmer who did not want to share his knowledge.

The French botanist Pierre-Marie-Alexis Millardet was visiting a farmer's field when he noticed that some of the grapes in the field, those nearest the road, were covered with a green powder. The grapes with the green powder did not seem to be suffering from powdery mildew. Millardet asked the farmer what the powder was, and the farmer pointed out what he thought would be obvious — that it was copper sulfate, a compound chosen so as to be innocuous to the grapes but dangerous-looking to passersby, encouraging them to leave the grapes alone. The farmer was either disliked by his neighbors, paranoid, or simply had grapes too good to resist. But what Millardet pointed out to the farmer was that the copper sulfate seemed to actually prevent powdery mildew.[22] Millardet went on to do years of research concerning the perfect way to apply copper sulfate to powdery mildew (with a heath broom dipped into either a bucket or watering pot filled with the compound). Copper sulfate also killed late blight, as Morren had anticipated.[23] Yet even after this was demonstrated, the use of the compound did not become widespread in Ireland until after World War I. Among other persistent challenges, it remained expensive and (even with a heath broom) hard to apply.

Since the end of World War I, a suite of new pathogens and pests that affect potatoes have emerged. Beasts with colorful names. Wart. Blackleg. Leafroll. Fusarium wilt. What's more, farmers and governments still spend an estimated $6 billion per year on the control of late blight, $1 billion of which is spent on fungicides alone, fungicides that in wet regions are sprayed on potatoes as frequently as ten times a year. The blight continues to evolve and threaten the world's potato crops. It is still a global danger whose consequences we will suffer anew if we fail to learn from the potato famine. But it isn't just the late blight of potatoes; thousands of pests and pathogens, along with changing climates, threaten our crops. To save ourselves from these species we must learn not only the story of the potato famine but also why the late blight that caused it was able to so quickly destroy all the potatoes in Ireland and much of northern Europe. Why was the destruction caused by that particular blight so pervasive? And are any of our modern crops similarly at risk?

3

The Perfect Pathological Storm

The potato fueled the rise of the West.
—Charles C. Mann, *1493: Uncovering
the New World Columbus Created*

A perfect storm results when the confluence of several phenomena turns a bad situation, sometimes a literal storm, into something far worse. The potato famine resulted from a perfect storm of
poor choices in addition to poor communication and major gaps in
our scientific understanding of pathogens. Historians, working with
biologists, finally have a sense of which phenomena made the late
blight so awful during the potato famine. Few of us, however, have
listened to the historians. As a result, rather than grow our crops in
ways that make disasters like the potato famine less likely, we have
done everything necessary to make such catastrophes more likely. A
perfect pathological storm gathers steam just over the horizon, and
rather than threatening only a few boats, it threatens whole countries.

Many famines had occurred before the 1840s, but never one like
the potato famine, never a famine of such great consequence tied so
directly to a single pathogen and a single crop. The human toll was
so great, in part, because of the extreme dependence of the Irish on
the potato. In this regard, many populations are as at risk today as
the Irish were then. But why did the pathogen kill so many potatoes?
Why were there none that seemed to survive? The answer is important because it bears on our modern agriculture; it speaks to the risk

our crops—including our modern potato—face today, including a recurrence of late blight.

In addition, the answer depends on decisions made long before the Irish ever started to grow potatoes, decisions made during the travels of the Spanish conquistadors and their aftermath. We left it to guys like Francisco Pizarro to choose the crops we now farm. He and other conquistadors may seem both repugnant and far removed from your daily life. Nonetheless, they influence nearly every bite of food.

Francisco Pizarro was born in Trujillo, Spain, around 1475, the illegitimate son of an economically marginal family. Some say he was an ugly baby; certainly he would become an ugly man. He grew up desperate, illiterate, hungry, strong-bodied, and morally loose. When the opportunity arose to travel to the coast in search of work and adventure, he took it. Once there, Pizarro joined the military for several years. He traveled to Italy and fought bravely, terribly, or both, according to his biographers. Eventually he boarded a boat headed for the Americas, dreaming of the riches he had heard others talk about in the long hours he spent on ships. Leaving Spain behind, Pizarro crossed the sea. He and other conquistadors like him were perhaps not the people to whom Western society should have entrusted the task of choosing the varieties of crops that moved around the world. Nonetheless, that is just what happened.

Pizarro's first trip from Spain to the Americas was part of an attempt to establish a new colony in what is now Colombia (with Alonso de Ojeda, who first traveled to the Americas on Columbus's second expedition). The conquistadors and colonists who followed in their wake planted seeds from European plants. They let loose horses, cows, and pigs. They wanted to be kings—or leaders, anyway—of a new, tropical Europe. But the colony failed. Many of the colonists died. Houses faded back into jungle soil, and the cows, pigs, and horses ran loose and multiplied. So, too, did some of the crops. Thanks to this effort and others like it, Pizarro and his fellow conquistadors spread crops and domesticated animals in the Americas.

But this was just one half of the great exchange; the other half involved the crops that would come back to Europe. But first those crops had to be found. In 1513, Pizarro and the explorer Vasco Núñez de Balboa headed west across the South American continent, searching for a way through to the Pacific. Improbably, they found one. Pizarro went on another expedition in 1523, again across Panama. Many died, and nothing was discovered. Nine years later, in 1532, Pizarro tried again, this time with his friend Diego de Almagro. They made it to the Pacific and again pressed on, following the still-unmapped coast. No one knew for sure what lay ahead, but there were stories. Men spoke of a great empire to the south, an empire of gold and riches. To arrive at the empire, one needed only follow the coast and then ascend the mountains. The coast would prove treacherous, and "the mountains [the Andes] were higher, the nights colder, the days hotter, the valleys deeper, the deserts drier, the distances longer"[1] than anywhere the conquistadors had been before.

What followed was the discovery of the Inca Empire, the death of the Inca ruler Atahualpa, the marriage of Francisco Pizarro to Atahualpa's sister, the birth of their daughter Francisca Pizarro (who would go on to lead an interesting life in Spain and to marry Pizarro's brother), the death of Francisco Pizarro at the hands of men loyal to the son of Pizarro's friend Almagro (Almagro himself was by then already dead; Pizarro killed him), and the removal to Spain of a great deal of Inca gold and even more silver. Also—and this is a big "also," perhaps the very biggest "also" of modern, Western, civilization— amid all this, the conquistadors moved crops from one place to another. As they did, the future of agriculture and humanity changed.

But which species would the conquistadors carry back to Europe from the temperate parts of the Americas? And to Africa and Asia from the tropical reaches? There were no easy answers. The decisions these men made about what to bring back from their travels affected the choice of crops from the Americas that we eat today; their choices lurk in the varieties of crops you find in the store, varieties that, as often as not, are the ones they picked. The conquistadors chose from among the plenty, but without regard for the centuries to come. In an ideal world, conquistadors such as Pizarro would have brought back many varieties of each species of the new crops they were encountering. These would have included varieties that differed in taste, in the climates and soils in which they might grow, and, as important as anything, in their resistance to pathogens. Of course what happened was the opposite. Consider, for example, the root crops of the Andes. At the time Pizarro arrived in the Andes the Inca farmed no fewer than ten thousand varieties of a dozen species of root crops. Pizarro and his men would have eaten many of these, cooked for them by their Native American wives. Of these multitudes a small subset was gathered by the Europeans. Perhaps one in ten thousand of these varieties made its way back to Europe. This raises two questions. First, why were so few varieties and crops brought back to Europe? Second, how were the varieties that made it back chosen? It is the answer to the latter question that was to shape the fate of the potato.

As to why so few varieties were brought back to Europe, the first problem was the conquistadors themselves. The conquistadors were not, for the most part, farmers. Nor were they skilled, necessarily, in learning from the locals. They were not even that good at distinguishing food from nonfood, much less the subtle differences among the former.[2] In addition, some species and varieties they did not see. No record seems to exist, for example, of encounters between conquistadors and the root crop oca (as a result, you might not have heard of oca). Other species they encountered but failed to note as food. Others still were recorded as food but viewed as unappetizing. Native Americans consumed frogs, beetles, termite queens, moths, bees, spiders, locusts, worms, mice, ticks, and algae. But because the conquistadors thought these items too strange to bring home, they, for the most part, never made it to our modern plates.[3] Ecologists talk about ecological filters, those features of habitats or of particular moments in evolutionary history that allow some species to move and thrive and prohibit others from doing so. The first crop filters all related to choices made by the conquistadors.

The conquistadors' preferences had lasting impact. If a food from the Americas was not tasty to the conquistadors, you are very unlikely to have ever seen it in a major store.

But even once the conquistadors decided to gather a particular plant or animal, whether it made it back across the ocean — or even to the coast—was another story, one in which ecology's laws were once again at play. The journey was long and terrible, especially for a delicate seed. When the men with whom Pizarro conquered the Inca Empire returned from the Andes, the route had many steps. First they had to descend the Andes to Arica, on the Chilean coast. In Arica they would board their boats and travel north to the Pacific coast of Colombia or Panama, then cross back through the jungles of the Isthmus of Panama to the Caribbean coast (the Panama Canal, of course, did not yet exist). Any seed or fruit traveling with them would have been exposed to constant humidity and months of conditions under which the most likely outcome was rot. It would have had to travel for miles in a small satchel roped to a dirty sailor or in a bag that was left banging on the back of a slowly dying mule in a mule train. Just a small handful of plants, whether in the form of roots, tubers, or seeds, made it all the way to the coast.

And that was just the first step. Once a plant made it to the Caribbean coast it still had to make it back in the ship; it had to survive the nearly four months it would take to get to Spain. On the ships of the earliest conquistadors, space was tight. Waste, rot, food, seeds, and sailors coexisted side by side, or failed to. The hold stank of excrement and rotten food. For reasons it is hard to imagine, given that the men were surrounded by the sea, the sailors sometimes threw the bones and offal of animals they ate into the bottom of the ship rather than overboard.[4] Rats often numbered, even on relatively small ships, in the thousands—a dozen rats per sailor, by one quite reasonable estimate. Shipwrecks from the time are riddled with gnaw marks made by the teeth of rodents. The conditions of the food on Columbus's fourth voyage to the Americas, for example, are said to have been so poor in hygiene and rich with life—grains and meats writhing with larvae—that the men preferred to eat at night so as to not see their food, not see what moved in their bowls. One ship from the late 1500s recovered off the coast of Florida contained evidence of three kinds of grain weevils, cockroaches, and dermestid beetles, all riding with the sailors from Europe to the Americas. The trip home would have been worse. And those same rats and writhing insects would eat any seed or root they came across.

Figure 3. Satellite image of the Canary Islands and their location relative to Africa. These islands, isolated and unusual though they might be, have played an outsize role in the history of agriculture. *Image by Jacques Descloitres, MODIS Rapid Response Team, NASA/GSFC.*

. . .

The entire trip from the Andes back to Spain took, on average, two years.[5] That any plant made it through these journeys alive is shocking. Those that made it did so because they were cared for and because they were the very hardiest and most indefatigable of the crops of the Americas. And even once a plant made it across the ocean, its journey into the European agricultural system was not complete. It still had to be able to grow somewhere in Europe. Nearly every trip to the Americas and back stopped in the Canary Islands (just as nearly every trip to Africa stopped in São Tomé and other islands off the coast of West Africa). The Canary Islands are volcanic subtropical islands. As a result, they include many of the key climates present in western Europe, including deserts, subtropical rain forests, temperate forests, grasslands, and even habitats that are snow-covered for part of the year. Crops could be established on the island as a kind of way station to further introduction. And whereas one had to travel among countries if one wanted to test out a potato in every potential climate in Europe, all one had to do in the biggest Canary Island, Tenerife, was travel from the coast to the top of the volcanic ridge. It is not surprising in this light that many crops that moved among continents did so via these islands—both from the Americas to Europe and from other parts of the world to the Canaries and then to the Americas. Banana and sugarcane (both native to the Far East), for example, spread to the Americas via the Canary Islands. Europeans then chose among crops of the Americas as they were coming from the Canary Islands. The Canary Islands became both a garden of possibilities and the place where all history associated with each crop was erased.[6]

The final step in filtering, intentionally or otherwise, the arriving crops into the subset we now eat was the process that occurred among European farmers and consumers. For sustenance crops, farmers nearly always chose the most fecund of the cultivars that arrived. This happened both because farmers preferred such varieties and because, once they began to grow them, the fecund varieties were those that a farmer was most likely to have in abundance. Each of these steps, these winnowings in the process of moving crops to Europe, could take whole human generations, but they rarely did.

Just thirty years after Pizarro raced up the hill to see the Inca Empire, just thirty years after he was bathed in silver, potatoes were being sold from the Canary Islands to mainland Spain.[7] Considering all the steps involved, that is as fast as potatoes could possibly become a commercial crop in Europe. We don't know how many varieties arrived initially, but we can surmise, based on the ecological filters through which they passed, that these varieties shared certain features. They were all able to survive on ships. They all arrived devoid of genetic diversity. The crops were also devoid (largely) of any partners they might have relied on in their native range. If their roots needed special fungi that helped them to access resources, for example, those were unlikely to have made the journey. If their flowers needed special pollinators, those, too, were left behind.[8]

The end result was that of the twenty-five root and tuber crops grown in South America, just the potato made it to the Canary Islands. Of the thousands of varieties of Andean potatoes (from nine separate subspecies), just a few dozen arrived in the Canary Islands, all of which appear to have been of the same subspecies (*Solanum tuberosum* ssp. tuberosum). Of the few dozen varieties of potatoes that made it to the Canary Islands, just a handful made it to continental Europe. Of that handful, just the lumper and a few others grew well in Ireland, where growing seasons are short and days during those seasons are long. Of the particular lumper lineages that were present in Ireland, those that were favored were either resistant to stress associated with transit, fecund, or able to grow where the season was short and the days long. The result was a crop that was fecund yet homogeneous, a crop that grew well but only in the absence of pests and pathogens. And there was something else, too.

The potato plants in Europe came ashore "naked," that is, without any of the traditional knowledge farmers in the Andes had acquired regarding planting, growing, storing, and preparing them over the course of centuries. "The complexity of Andean cropping systems had no precedent in Europe," says James Lang, author of *Notes of a Potato Watcher*. It was "geared to every nuance of altitude and rainfall." In other words, the conquistadors had a lot of catching up to do.[9] They could have learned a great deal that they could have passed on to people farming potatoes in the Canary Islands, who could have in turn passed the knowledge to those who began to farm potatoes in Europe. But this did not happen. As a result, by the time the late blight was moving across continental Europe and then to England and Ireland, nearly all the potatoes were of a single highly productive variety, the lumper, and they were being farmed in new ways, recently invented by Europeans, invented without the biology of the potato or its pathogens very much in mind.[10]

Everything the British government and agricultural scientists urged Irish potato farmers to do, and nearly all the choices the farmers themselves made, sped up the spread of late blight. First, the potatoes were planted as monocultures, on which the Irish depended exclusively for their sustenance. We tell ourselves, "Don't put all your eggs in one basket or all your fields in one crop," yet we do anyway. Second, the British urged the Irish to abandon their traditional raised-field planting technique and instead to plow their fields. This made conditions better for late blight, inasmuch as the raised fields (akin to those used in the Andes) increased temperatures enough to kill the pathogen. The flat, plowed fields, on the other hand, made life easier for late blight.

In addition, Andeans replanted their fields with seed potatoes, but they also paid attention to and used those potatoes that resulted from true seeds. The seeds of potatoes are harder to work with. They are unpredictable (the seed of a potato can produce a potato very different from the one out of which it has grown, thanks to the genes carried in pollen).[11] Yet Andean farmers knew (and know) to keep an eye out for those occasions when true potato seeds yield, at the end of fields or elsewhere, varieties that taste better, grow faster, or, in this case, don't die from a pathogen killing everything else. The Irish and other Europeans, by contrast, replanted each year exclusively using hunks of potatoes, hunks that are confusingly called seed potatoes. Cultivating a seed potato is the sort of thing you can do as an after-school project using a glass of water and toothpicks. Choose an old potato from your cabinet, one with a sprouting eye. Replant it. That will be enough to restock your potatoes forever, so long as conditions are good and unchanging.

In some cases Irish farmers didn't seem to have known that potato seeds could be used to grow a potato. Of course using seed potatoes ensured that the following year's potatoes would be clones of those that were originally planted (and for this reason, Andeans also mostly plant seed potatoes). If the first generation of potatoes grows fast, this is great. But an advantage in the short term can be a disadvantage in the long term. The Irish had not, until the arrival of late blight, had a chance to learn about that disadvantage. As a result, nearly all the potatoes in Ireland were genetically identical to each other and genetically identical to every other potato that had ever been planted in Ireland. To the extent that any differences existed, they were attributable to chance mutations, most of which tended to be deleterious.

In many cases what results after a tragedy like the Irish potato famine is that, very quickly, any lessons that might be learned are forgotten by most people. Tragedies of our ancestors seem remote to our ordinary lives. Often, though, a few individuals (or institutions) remember and learn. The progress of civilization and our hope that it continues depends disproportionately on these few. For example, while most of the world all but ignored the traditional knowledge held by the Andeans about potatoes during the potato famine, after the potato famine and through to today, a few scholars paid attention. Their work has begun to pay off, which is important not only because potatoes suffer from many different problems but also because the late blight of potatoes never went away. Scientists have started to examine whether any of the Andean potato varieties are resistant to late blight. This is something you might imagine would have happened in the late 1800s. It didn't. It is just happening now, and it is only possible because of an amazing act of foresight almost half a century earlier—an act of foresight, luck, and, to some extent, a reliable truck.

. . .

In 1971, the Peruvian government created the International Potato Center (CIP, or Centro Internacional de la Papa). One goal of the center was (and is) to study and conserve the traditional varieties of the potato and other crops of Peru for the people of Peru and for the world. It was to be (and is) one of eleven such centers, each of which is dedicated to a different sort of crop in a different region; the centers are loosely coordinated by CGIAR, the Consultative Group for International Agricultural Research.[12] By 1982, the center's collection was large: it included thousands of varieties of potatoes, not to mention other crops. It was successful. It was also in jeopardy.

In 1982, the country was in the midst of a guerrilla-led civil war in which the Marxist guerrillas, the Sendero Luminoso (Shining Path), made daily life both difficult and terrifying for many Peruvians. For example, the Sendero Luminoso arrived at one of the main agricultural stations where native Andean crops, including potatoes, were being farmed and conserved, in Ayacucho. They surrounded the station bearing lit torches and prepared to burn it down.[13] The scientists feared for their lives and ran. But a campesino, a poor farmer, stepped forward. He begged the Sendero Luminoso to leave the collection be, not because the scientists needed the seeds and potatoes but because the farmers, the people, needed them. It worked. Or at least it bought the scientists time.

According to James Lang, in *Notes of a Potato Watcher,* Carlos Arbizu, an agronomist employed at the time by the National University in Ayacucho, decided he must move the collection. At the urging of the head of the station, he packed as much of it as he could into his truck and drove off. The next night, the Sendero Luminoso destroyed the building in which the seeds, tubers, and roots had been stored. Arbizu was still driving when it happened. In Lang's telling, Arbizu asked people, in any village in the high Andes in which he could safely stop, to plant and care for some of the tubers and seeds. Later, Arbizu went back to the villages to collect some of the samples for the nation. The farmers had kept farming many of the varieties that they liked and, in doing so, helped save them.

Today the International Potato Center has expanded its collection and mission far beyond what might have seemed possible in the hard years of the Sendero Luminoso. The center saves and studies potatoes of nine different subspecies, some of them cones, some crescents; some red, some blue, some purple; some rich in protein, others in vitamin C. The center also works to save other traditional Andean root and tuber crops that never made it back with Pizarro or other conquistadors. Oca. Mashua. Ulluco. Maca. Arracacha. Mauka. Ahipa. Yacón. Achira. Each one neglected, threatened, and special in some way. The center also continues to try to find lost varieties of traditional crops and to study and use and make available the unique values of those varieties, whether they be flavors or types of resistance. Until recently, the consensus in scientific literature was that no known potato variety could resist potato blight. But that consensus turned out to be premature. In fact Andean potato varieties differ from one another in nearly every attribute, including their resistance to blight. This is not surprising; it just wasn't known.

In 2014, Willmer Pérez, a scientist at the International Potato Center, started to check traditional potato varieties for their resistance to late blight. Nineteen of the 468 varieties tested, distributed across seven species, were highly resistant to at least one form of blight.[14] Pérez has yet to check the many hundreds of varieties in the potato center's collection, much less the more than four thousand varieties of potatoes grown in the Andean highlands of Peru, Bolivia, and Ecuador. He hasn't checked them for resistance to blight, nor has anyone checked most of them for resistance to fungal, bacterial, or viral pathogens or insect pests for that matter. Pérez has struggled to find money to support this work and can only do so much on his own. Too many potatoes, too little time.[15]

Meanwhile, farms in Ireland still plant just the handful of potato varieties brought back by the conquistadors, none of which is resistant to blight. It took just a few decades for Pizarro and other conquistadors to travel from Spain to the Inca Empire, conquer it, and bring back the potato. It has taken nearly five hundred years, though, to appreciate the real treasures of that empire—its wild diversity of crops and its inhabitants' knowledge of them.[16] Today, our knowledge is out in the world, as was Berkeley's during the famine. It has been published by scientists, but it is as of yet without consequence. Failures of communication persist.

If we were to live through the potato famine again, one would hope that we would be quicker to learn from new scientific studies of crops and their enemies as well as from ancient traditional knowledge. Of course the easiest solution is simply to avoid moving the pests and pathogens that affect crops around the world. Indeed, having benefited from moving crops far away from their pests and pathogens, we are rather careless about reuniting them. As the writer David Quammen has observed, "Everything, including pestilence, comes from somewhere." The blight came from somewhere, and knowledge of its source can (or might, anyway) help prevent the recurrence of blight in the future. Until the work of Tom Gilbert and Jean Ristaino was published, we didn't have that knowledge.

Tom Gilbert is a clever British boy wonder of a scientist who is often described (not entirely truthfully) as the youngest full professor in Europe. Tom works at the Natural History Museum of Denmark, where I work during the summers. There he specializes in coming up with technically challenging but novel approaches to dealing with hard problems—the kind of problem doesn't really matter, so long as genes are involved. Tom hates easy problems. Hard problems, on the other hand, make him giddy, or at least giddy in a dry, British kind of way. Among other things, he has attempted to sequence the genes of the extinct great auk (successfully, Tom would add if he were here). Other projects involve giant squid, the evolution of pigeons (which entailed a lot of pigeon shooting in the middle of European cities), and an attempt to figure out the kind of wine a person is drinking based on the DNA present in the bottle (which would allow counterfeit wines to be identified). It is this latter project that explains, I assume, why the last time I visited Tom's office I saw a cooler marked ROYAL WINE, in which one could find hundred-year-old bottles from the cellar of Queen Margrethe. A visit to Tom's office is likely to yield a cup of coffee, a very good conversation, and an interruption by some student who stumbles in, eager to ask a question about, say, the sample of vampire bat blood she holds in her hand.

Jean Ristaino, on the other hand, is a plant pathologist. Plant pathologists study the fungi, oomycetes, viruses, and bacteria that kill plants. The direct intellectual descendants of Miles Berkeley and Heinrich Anton de Bary, they use every tool available to identify pathogens. Their work is what saves our crops from destruction again and again—or fails to. And yet because their work is not viewed as sexy, or the next big thing, plant pathologists have become increasingly rare. Even as universities get bigger, they tend to have more biologists who focus on applying a method cleverly and fewer plant pathologists and other biologists skilled in knowing an organism well (more Toms and fewer Jeans).

Jean Ristaino has spent her career studying potato blight. She desperately wants to understand it, not just because of its important history but also because we still have not escaped it. Among the most vexing aspects of the story of blight has been the question of where it came from and whether the blight we are dealing with today is the same one that was present in the 1840s in Ireland and the rest of Europe. In theory, one could compare samples of the two, but in practice so little biology was being done during the nineteenth-century blight that few samples were taken. But Ristaino persevered and found some samples in European and US herbaria (collections of dried plants). In 2001, she showed that she could find DNA in the old samples and use it to identify the strain of the late blight of 1845, which she revealed was not the same as the one then present (in 2001) in Ireland or the one that was widespread globally in the mid-twentieth century.[17] The next step was to consider the old blight's code in detail, and that's where Tom Gilbert came in. After seeing Ristaino's announcement that she had found blight DNA from the old samples, he enlisted someone in his lab to decode it. That someone was Mike Martin. Here, in this collaboration between Tom, Mike, and Jean, lay the great hope for understanding late blight as well as the sort of approach that might help us to make sense of the many species that threaten our crops.

When Mike and Tom found and decoded the DNA in the old samples, they and Jean were in for several surprises. First, the late blight that caused the Irish potato famine appeared to be one most closely related to blights found today in the Andes. (Others had suggested Mexico; a minor, sometimes nasty, war persists as to who is right. The war's resolution is tipping toward the Andes, or at least it seems so from my perspective, which is influenced by my conversations with Jean and Tom.) This suggested that the blight was an ancient adversary of potatoes. It also suggested that in the 1840s someone unintentionally brought the blight over from the Andes into an environment where the potato was far less protected than it would ever have been in its native land (first through movement to the ports of New York and Philadelphia and then on to Belgium). Some people have suggested that this dispersal occurred when potato biologists were trying to find less degenerate seed potatoes in the years leading up the appearance of the late blight in Europe. Others suggest that it occurred when a few potatoes traveled in a shipment of guano (fertilizer) from the Peruvian coast. Perhaps the truth lies in some mix of the two.

. . .

Subsequent studies have revealed more surprises. First, the strain of late blight from 1845 is not extinct.[18] It can still be found in both Mexico and Ecuador, lurking like the ghost of horrors past.[19] Second, the blight currently found in Ireland is not the same strain. Third, the blight currently found in Ireland is diverse; it is not a blight but rather several blights, which appear to have been introduced to Ireland *after* the potato famine. Even after more than one and a half million people died as a result of the potato blight, we still kept introducing new kinds of blight, even though we failed to introduce new kinds of potatoes. Worse than that, not only are the strains of late blight in both Europe and North America now diverse, they also include strains that resist some of our best fungicides. The new strains of blight include sexual and asexual forms. They even include newly evolved forms, some of which are more dangerous and aggressive than any potato blights ever seen before. The blight does not learn, but in response to natural selection (and with our actions as aid), it has done everything right to ensure its success. On behalf of the potato, we continue to do everything wrong.[20]

Each thing that happened to the potato in Ireland could happen to almost any of the plants we most depend on anywhere. It could happen because of an oomycete, a fungus, a virus, or an insect. When it does, we will depend on the knowledge—and actions—of scientists and other scholars.

For example, having contributed his ideas to the discussion of the blight, Miles Berkeley simply hoped that someone else would act upon those ideas, and his behavior suited the norms of the time. As decades have passed, the role of scientists in society has changed in some ways. Many scientists, even those focused on very basic problems, now view engaging policy makers and the public as part of their jobs. Land-grant institutions in the United States, such as the one at which I work, were founded in part to make just this connection. But nonetheless, the number of specialists trained to save a particular crop is, in almost all cases, very few. As a result, whether a crop is saved in the nick of time (or, far more rarely, well in advance) almost inevitably depends on the actions of just one or a small handful of individuals — poorly funded individuals, tired individuals, dedicated individuals who love crops or even pests and pathogens far more than reason suggests is normal. Historians dislike the "great person" version of history. Structures, policies, and social trends, they say, matter as much as does a single person or a small group of people. Certainly this was true in the case of the potato blight. One reason that the famine was so terrible had to do with the policies of the British and their belief that the free market could solve anything. But it is also true that had one or another scientist acted sufficiently decisively, things might have been different. The same has proved to be true in dozens of cases since. A great deal depends upon a tiny handful of scientists, and the situation, if anything, is getting worse. The number of pests and pathogens threatening crops is increasing faster than the number of specialists trained to fight them.

During the years between 1845 and today, new pathogens and pests have threatened every major crop. But in the years between 2006 and 2016, the rate of these new threats has accelerated. What used to be a slow tap, tap, tap on the thin ice of civilization has turned into a pounding, drumming announcement. And just as the potato blight, in retrospect, has specific causes, many of them preventable, so do the new, dangerous pathogens and pests on the rise among crops today. Having failed to learn crucial lessons from the potato famine, we were doomed to repeatedly revisit our errors (and, when we do, to hope that someone saves us from our own mistakes in time). But we can also see them with greater clarity as consequences of basic laws of nature. Among those laws is the law of area, which says that the larger the area planted, the more likely it is that a crop will be colonized by ever-more-novel pests. In Africa, no crop offers a bigger target than cassava, a crop on which hundreds of millions of people depend, just as the Irish once depended on the potato.

4

Escape Is Temporary

We may infer from these facts, what havoc the introduction of any new beast of prey must cause in a country, before the instincts of the indigenous inhabitants have become adapted to the stranger's craft or power.

—Charles Darwin, *The Voyage of the Beagle*

We worry about pathogens that attack humans, such as Ebola, MERS (Middle East respiratory syndrome), and Zika. But if the potato famine teaches us anything, it is that we should also worry about the dangers to our crops. They are inevitable.

In 1970, a malady affecting cassava arrived in the People's Republic of the Congo (later Zaire) and the Democratic Republic of the Congo. We don't know for sure when it was first noticed. A farmer likely went to the cassava growing in the field behind her house, the cassava that was to feed her family, and saw that the plants were sick. The stalks and leaves lay black and twisted on the ground. When she got back to her village she found that other women, too, had gone to their fields to collect cassava. They also saw the sickness. The stems of the cassava plants were stunted and weak, their leaves distorted, their roots—the foodstuff—small. The women pulled the diseased cassava out of the ground and planted more of the same. The new plants also suffered.

What was causing this destruction? As farmers talked about the problems, they found a culprit, a small insect on the plants with the face of a monkey, complete with large "nostrils." It was an emissary of some lesser and wicked god. This monkey-faced beast was to blame![1] Moreover, it seemed to be a kind of a messenger of misfortune, a bad omen. As a result, no one would touch the animal. Meanwhile, the malady spread.[2]

Technicians at an experimental farm in Mantsoumba, set up by the government twenty kilometers (twelve or so miles) west of Brazzaville,[3] had seen the damage. They went out to their own fields and looked carefully at their plants. They noted the monkey-faced insect. It would prove to be the chrysalis of a butterfly (*Spalgis lemolea*), a chrysalis that is said to mimic a tiny monkey head so as not to get eaten by birds. It was not the cause of the malady. Instead the technicians observed sucking insects of diverse sorts on the cassava plants, including a species they had not seen before.[4] The new species appeared to be the worst among them. It was small and white, an animal so modest and featureless as to scarcely deserve being called an animal. It looked to be a kind of mealybug.[5]

The details of what happened next are muddied by time and conflicting accounts. An agricultural engineer at the ministry of agriculture in the People's Republic of the Congo and a professor of zoology at the University of Brazzaville became involved. In 1973, they or someone who worked with them sent a sample of the mealybugs plaguing the cassava to the National Museum of Natural History in Paris, where it landed on the desk of Danièle Matile-Ferrero,[6] one of relatively few world experts in mealybugs. Matile-Ferrero prepared the specimen carefully. She then looked through a microscope at its features. Some attributes of insects can easily be seen through a microscope. Others must be inferred based on intuition developed over years of experience. Matile-Ferrero formed an image in her head of the specimen before her. She made a drawing. She compared the image and the drawing to other species in her collection and in every other key collection around the world. Slowly, meticulously, she was coming to the conclusion that the animal before her was totally new to science and that she would need to give it a name.

. . .

Cassava is an unglamorous food. Like the potato and other roots and tubers of the tropics, its story is one of food as energy rather than luxury.[7] Like the potato, its taste is humble. Like the potato, it offers nearly complete nutrition. You can live off it with little else as a supplement if you eat both its leaves and its tubers. Whatever it lacks in subtlety and flavor it makes up for in fecundity. Plant the sticks in poor soil, and the plant will grow and make large storage roots, some weighing as much as ten pounds. Harvest one and another will grow. Come back several weeks later and there will be more.

Cassava is not native to Africa; it comes from the Americas. The conquistadors saw it, ate it, and left it where they found it. Tropical food plants were not useful to colonial powers unless they could be farmed in the relatively cool, dry conditions of Europe. The exceptions were luxury crops such as cacao, tobacco, sugarcane, and coffee, which could be farmed in the tropics and then exported to Europe. Cassava was no luxury. Yet four hundred years ago Portuguese traders collected cassava stems from Brazil and carried them first to the islands off West Africa, including São Tomé, then to the Congo River delta of Central Africa,[8] where they were planted, then to Asia, where they were planted again. Once on these new continents, cassava spread from village to village, playing a bigger role each time there was a bad season, a bad year. The more frequent hardships were, the more quickly cassava spread, until many became dependent upon it. Just as with potatoes in Ireland. Just as with sweet potatoes in China. Cassava is the primary source of calories in sub-Saharan Africa, the primary source of calories for five hundred million people. This is particularly true in Central African countries such as the Democratic Republic of the Congo, where as much as 80 percent of all daily energy comes from it—a single type of plant. In the DRC, one brings cassava leaves when visiting a house;[9] the plant is a gift synonymous with life.

If a pest or pathogen were ever to destroy cassava plants across Africa the way that late blight destroyed potatoes across Ireland, it would be a desperate tragedy. It would be a tragedy for the hundreds of millions of children and adults whose skin sweats with its scent. In sub-Saharan Africa alone, nearly ten million hectares are planted in cassava. Cassava is too important to collapse, especially as soils get worse and climates more hostile, conditions in which few other crops will grow. Cassava is the secret strength and weakness of Africa, tropical Asia, and the world. But whereas hundreds of millions of dollars are spent studying coffee and wine, the funds available to study and monitor cassava have always been modest. They always will be. Cassava lives in the empires of dirt, where the poor raise civilizations out of red mud.

As a result, one thing was clear even before Danièle Matile-Ferrero identified the mealybug that had been sent to her: the Congo basin, and potentially much of Africa and Asia, had a big problem—and since the mealybug appeared to be new to science, it was a problem about which essentially nothing was known. Mealybugs do not look ferocious. They are delicate and seemingly without skeleton, a smudge of life. But what research would soon make clear is that this particular mealybug has a special fondness for the food Africa depends on and an ability to reproduce with sufficient speed to decimate all the cassava in Africa and, with it, millions of Africans. Tiny, soft, harmless-seeming mealybugs have the power to destabilize a whole continent. In the villages in which the mealybug had arrived, it was an affliction tough to distinguish from fate. Little could be done. Then, as now, few could afford pesticide.

In France, Danièle Matile-Ferrero continued to study the mealybug she had been sent. Then in 1977, no fewer than five years after it was first detected, she named it *Phenacoccus manihoti*—the cassava mealybug. It took five years for the insect threatening the main foodstuff of much of Africa to be named. The question was, what next? What could be done now that the insect had been identified? Matile-Ferrero was invited to the Congo basin to study the insect in more detail. Once there, she found that it had spread widely and was destroying whole fields of crops. She recommended controlling the spread of the mealybug through three methods: the targeted use of pesticides to treat any cassava that was being moved (which was cheaper than spraying whole fields), the breeding of resistant varieties, and biological control—the use of the natural enemies of a pest—its predators, pathogens, and parasites—to control its populations. Ironically, the monkey-beast that was initially accused of killing the plants was, as a caterpillar, a predator of mealybugs and hence one potential agent of biological control, if it could be bred in a lab and released in large numbers. But the real hope, it seemed, lay in finding another insect that could more effectively eat the mealybugs, a superagent of biological control. She listed some possibilities, insects she had collected that eat mealybugs in the field, but she noted that they seemed relatively ineffective and that they would need to be sent to other experts for more study.[10]

. . .

One person who would follow up on Matile-Ferrero's suggestions, Hans Herren, was, at the time, still thousands of miles away. He was a young Swiss hippie working at the University of California at Berkeley, steeped in the ethos of a town where even the marijuana is organic. Herren grew up in Switzerland on a farm in the lower Rhone valley, where his father grew tobacco, wheat, and potatoes using traditional farming techniques. Under the wheat, his father grew clover. The clover shaded out weeds and could be sold to farmers. Other farmers, in turn, would provide Herren's father with manure for the fields. Herren's father bought little in the way of pesticides and fertilizer. Then one day, Herren remembers, a big black car pulled up from Basel. Men from the city had come to try to persuade his father to use fertilizer and pesticides, to modernize.[11] After that, as Herren remembers it, the yields on the farm were higher, but so were the costs—the costs of fertilizer, pesticide, and gas for the tractor and the costs to the environment. It was the distinction between farming before the men in the black car pulled up and after that led Herren to think about the benefits of working with nature rather than against it.

Once Herren finished his PhD, he considered taking a pest-control job in Switzerland, but he suspected that such a job would focus on chemistry—on better ways to kill insects using pesticides. Herren wanted something else. He applied for and got a postdoctoral position at the University of California to work with Robert van den Bosch, who was himself the son of a Swiss mother who grew up on a farm. Van den Bosch (Van, to those who knew him) was in the process of writing a book critical of pesticide companies.[12] These companies were, in turn, threatening him with lawsuits. Then, not long after his book was published, van den Bosch died of a heart attack. Inspired by Van's work and saddened by his loss, Herren was determined to do meaningful work in the world. Then he saw an advertisement for a job in Nigeria with the International Institute of Tropical Agriculture (IITA): they were looking for a plant breeder to work with corn (or maize, as it is known outside the United States). It wasn't what he was trained for, but it seemed like an adventure, and he had the necessary skills.

What Herren could not have known was that his application had landed on the desk of Bill Gamble, at the time the director general of the IITA. Seeing Herren's résumé, Gamble decided he needed Herren not for the job for which he had applied but for another. He needed him, with his background in crops and insects, to help try to figure out how to control the cassava mealybug. The two men met, and Gamble offered Herren the job on the spot. Herren soon left for Nigeria with big aspirations. He would stop the cassava mealybug and, in doing so, save a continent.

By the time Hans Herren arrived in Nigeria, where the IITA office was located, the spread of the cassava mealybug was extensive. Since 1973, the mealybug had moved from field to field in the Congo basin and then began to arrive in countries as far away as Senegal. It was spreading from port city to port city, perhaps as folks traveled to see their friends, bringing with them bundles of cassava leaves as gifts. How long ago this spread had occurred was hard to say, as was just how common it was farther out in the bush. When a new pathogen or pest is found on a crop, it has typically been around for a while. Farmers rarely report new encounters until the invader has become a problem. In this case, farmers may have been slow to report, but what took even more time was for people to pay attention to the farmers' pleas for help. The mealybug had spread without challenge through the unprotected fields; it could spread in the gift of leaves. It could even spread, it would later be learned, by floating up into the wind.

As Hans Herren considered the mealybug and how to control it, he read Matile-Ferrero's papers. He paid special attention to her notes about the likely origin of the mealybug. The insect, she noted, was most closely related to other species from the Americas. On the basis of this information, she hypothesized that it was from the native range of cassava and was succeeding in West Africa in part because it had found its food and escaped its own threats. Relatively unhindered by predators and parasites (the monkey-beast notwithstanding), the mealybugs ate—or, rather, sucked—with impunity. They sucked and mated and sucked some more. But just where the cassava mealybug came from—that was anyone's guess. "The Americas" was all Matile-Ferrero could determine with any certainty.

Cassava is a domesticated plant. Relative to their ancient, wild ancestors, the domesticated plants on which we depend are often defenseless.[13] Author Annie Dillard has written at length about Galápagos Islands animals that do not seem to know predators. They came right up to her, she says—the tortoises with their shells full of meat, the salt-spitting iguanas, the wingless cormorants, the penguins and the sea lions. Even the flies, she says, hover as if unafraid. They are naive, innocent, living in a bloodless land as if before the fall. Our crops are similar. Their innocence is less conspicuous than that of a flightless bird walking toward a large human yet far more consequential.

The relative innocence of our crops has two origins. We breed crops to produce as much food as possible, to grow without any concern other than getting as close to the sun as possible as fast as possible. This growth comes at the expense of the plants' ability to defend themselves, and the more industrialized agriculture becomes, the more this is true. In addition, in some cases the defenses we use to protect plants against herbivores are also toxic to us, so by making plants more productive and less well defended, we also make them more palatable. But there is also something else, what scientists call enemy release. We move those crops to places where they will be safest, where they can most easily be released from the threats of their enemies and put all their energy into luxurious growth. Life is easier if you can escape the things that eat you, and one of the ways to escape them is to move or, if you are lucky enough, be moved. This geographic release from danger, from enemies, is central to understanding the threat to cassava as well as many other modern agricultural plants,[14] and once plants achieve it, they are even more likely to invest in growth over defense.

If one were to draw a map of where the major crop plants are farmed today and where they were first farmed and domesticated, an unusual geography emerges. Crops are usually domesticated in one place, often in mountainous regions. They are then shifted to a new place, often a river valley associated with the rise of one or another early civilization. Finally, with the globalization of the world, in the era of ships, they began their longest journeys, to continents where they are not native. The biological stories of our ancestors and their crops repeat themselves. Life, while diverse beyond our ability to measure, obeys a set of laws. We have not escaped these laws.

Potatoes, for example, were domesticated in Chile and the Andes but are now grown mostly in North America and Europe. Vanilla was domesticated in tropical Mexico and is now farmed in Madagascar, Indonesia, and China. Squash, pumpkins, and some gourds were domesticated in the Americas but are now primarily grown in China. Sweet potatoes were domesticated in Central and South America but are now farmed primarily in China. Bananas were domesticated in Papua New Guinea, but the bananas that are exported are grown mostly in Central and South America. Rubber (*Hevea brasiliensis*) is native to the Amazon, but nearly all commercial rubber is now grown in tropical Asia. The cacao tree (*Theobroma cacao*), whose beans are used to produce chocolate, is native to the Amazon, was domesticated in Mesoamerica, and is now mostly grown in West Africa. Similar patterns exist for most of our crops, especially those from the tropics. These geographies are all the result of enemy release, our temporarily successful creation of worlds for our plants in which they are free of enemies.[15] Yet just as predictable as this release is the inevitability that eventually the pests and pathogens will catch up. The job of border controls and quarantine offices is to slow down this catching up, to stall the herbivorous monsters, fungi, bacteria, and viruses in the absence of which our crops have thrived. Today, given the number of airline flights and ship crossings people undertake, this task is harder than it has ever been.[16]

Just how many pests and pathogens have yet to move? How charmed is our moment in history? In 2014, Sarah Gurr and her colleagues at the University of Exeter tallied all the pests and pathogens affecting the world's main crop species in the regions in which they are native and in each of the places where they have been introduced. They focused on 1,901 enemies for which data were available in the Plantwise database, developed by the Centre for Agriculture and Biosciences International. The data in Plantwise are incomplete. Some of the identifications of pests are likely wrong (many pests and pathogens do not yet have names, after all). And the identity of the host on which each enemy was found is often uncertain. An entry might indicate that a pathogen was found on wheat but not distinguish among the many thousands of different varieties of the plant. Yet these data represent the best understanding we have so far of the enemies of agriculture and their distribution.[17] Sarah's assumption was that each of these crops, upon initial introduction into a new realm, whenever that might have occurred, was devoid or nearly devoid of pests and pathogens. Then slowly, following accidental introductions on ships, on trains, and in cars, the beasts began to catch up. But how fast? And where do crops still have the greatest reprieve and hence the most to lose with further introductions?

The enemies, Gurr and her colleagues found, have all caught up to some crops. Other crops, though, a few, are still experiencing something akin to complete escape. They live in a world without dangers. Patterns emerged when Gurr considered which enemies have caught up and which have not. Small things—oomycetes, fungi, bacteria, and viruses—catch up first. Very few of these pathogens are still found only in the regions in which their host crops were domesticated. If a fungus is not yet in a particular place, the odds that it will get there quickly are high. But Gurr also found something else, a geographic pattern to where the pests had and had not caught up. The crops being farmed in affluent countries, including the

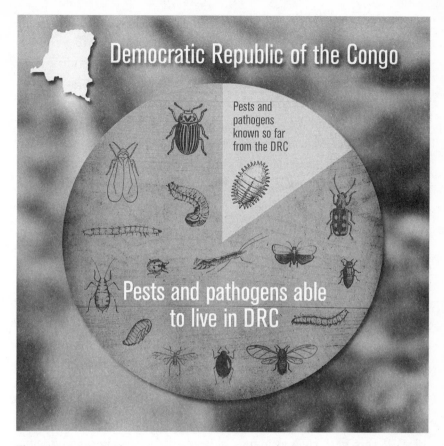

Figure 4. In tropical countries such as the Democratic Republic of the Congo, most crop pests and pathogens that could live in the country—an assessment based on the climatic conditions and the crops being grown—have yet to arrive. For now, these species are held back by good luck and the human control and inspection of products being shipped to these countries. *Data are drawn from Daniel P. Bebber, Timothy Holmes, and Sarah J. Gurr, "The Global Spread of Crop Pests and Pathogens," Global Ecology and Biogeography 23, no. 12 (December 2014): 1398–1407.*

United States, France, Italy, the United Kingdom, and Australia, and in some rapidly developing countries such as China and India, have already, for the most part, lost the advantage earned through escape. Most of the enemies of crops in these regions have caught up. At current rates of accidental introduction of enemies in these regions, they will have all arrived by 2050. For crops in these regions, the only new reprieve would come from moving them to another planet or moving them indoors. (In practice, we create fields free of enemies through pesticides, though this, too, has its limits, to which we will return.[18]) Conversely, in the tropics one finds many regions in which most pests have not yet caught up with their crops. On average, tropical countries still host just one-fifth of the crop enemies that could live there, were all crop parasites and pathogens to arrive everywhere their hosts live. One-fifth. The other way to put this, of course, is that four-fifths of the enemies that could live in any given tropical country have yet to arrive. We can't say when this will happen, but we can predict which crops it is likely to happen to first.[19]

.   .   .

Don Strong, now with the University of California at Davis, spent years considering which factors determine the rate of colonization of animal species in various habitats, including crop fields. He studied, for example, the pests affecting sugarcane and coffee plantations.[20] He was interested in general theories underlying the workings of life on earth. But sometimes the search for generality also offers us particulars. When Strong compared the number of kinds of pests in sugarcane fields around the world, that number was predicted by one variable: the size of the field. Bigger fields had more kinds of pests;[21] this pattern was a specific instance of a general phenomenon, the species-area rule. In 1977, when Strong carried out the study, the tiny island of Dominica, less than a third the size of Rhode Island, had around two hundred hectares of sugarcane and just six kinds of sugarcane pest. Much larger Puerto Rico had 151,000 hectares of sugarcane and more than a hundred kinds of sugarcane pests. A similar pattern emerges when one studies cacao plantations.[22] Big areas provide more food than small areas; they are also bigger targets for pests arriving from afar. Where crops are non-native, pests arrive both through adaptation to the new crops (evolution favors any organism that can feed on an abundant food) and through dispersal from the crop's motherlands. Subsequent research has shown that time also matters. The longer a crop has been planted in a particular area, the more likely the species that can colonize it will have done so.[23] For the theory of island biogeography, this is a beautiful illustration of the general rule that habitats with the largest amounts of area will contain the largest number of species. But for agriculture it has a sinister significance. It means that the more dependent we become on a crop, the more likely it is to be destroyed; this is what I am calling Strong's paradox. Our most precious foods are the most susceptible targets.

Combined with Gurr's results, Strong's paradox leads us to the ominous prediction that our enemies will all eventually catch us and that they will attack us first where it most hurts, in our biggest, most sustaining fields—our fields of cassava, for example. Such targets are humanity's weakness. If the mealybug hadn't shown up, something else would have. In fact, since the mealybug first appeared, nearly a dozen other pests that affect cassava have arrived in Africa. Given the area over which cassava is planted, it was only a matter of time before some pest arrived. That the pest would be the mealybug was up to chance or luck. Very bad luck, as it would turn out.

For the cassava mealybug the most likely scenario was that nothing would stop its spread across Africa and maybe even across Asia. And while the losses these mealybugs caused were variable—depending on climate, the rainfall in a particular year, and the variety of cassava planted—they were often high, sometimes complete. Whole farms of cassava were lost. Wherever they went, the mealybugs seemed likely to leave hunger in their wake as certainly as if they had attacked the humans themselves.

5

My Enemy's Enemy Is My Friend

Let a man profess to have discovered some new Patent Powder Pimperlimplimp, a single pinch of which being thrown into each corner of a field will kill every bug throughout its whole extent, and people will listen to him with attention and respect. But tell them of any simple common-sense plan, based upon correct scientific principles, to check and keep within reasonable bounds the insect foes of the farmer, and they will laugh you to scorn.
> —Benjamin Walsh, *The Practical Entomologist*

In 1958 Chairman Mao Tse-tung decided it would be a good idea to rid China of its pests. He stood, a plump and confident man, and pronounced the names of the animals he intended to exterminate. Fleas. Rats. Sparrows. Flies.

The idea of trying to kill off one or another species was not a new concept. Among the first uses for human tools was to get rid of species we didn't like (and to eat the ones we did). One can tell the entire story of humans based on the things we killed at various moments in our past. What was novel was that Mao decided to do it on one day, across an entire country the size of North America. Mao was going big.

Everyone in China was ordered to do whatever he or she could to kill these four kinds of animals. The efforts were most conspicuous with the sparrows. People were ordered to go outside and bang pots and pans for forty-eight hours straight (taking turns) so as to cause the sparrows to fly in fear until they were exhausted and died. Sparrow eggs were also to be found and destroyed. And anything flying could, and should, be shot, in case it was a sparrow.

It is hard to know just how many animals were killed during this two-day episode. Mao's government claimed quantities extraordinary in their magnitude and precision: 48,695.49 kilos of flies, 930,486 rats, and 1,367,440 individual sparrows. Whatever the true numbers, sparrows in particular were incredibly rare after this killing spree. Cities and countrysides went quiet. This was a victory for Mao in that his plan had been successfully carried out. Unfortunately, tree sparrows, the most common sparrow in China at the time, eat far more insects than grain and benefit crops by controlling pests. They are, or were, secret agents of biological control (a fact known by ornithologists but ignored at the time). Once the sparrows were gone, pests in crops reached outbreak levels. The crops began to fail. Had this been a problem in isolation, things might have been okay. But Mao made many other decisions that contributed to food scarcity. As a result, the greatest famine in human history ensued. At least thirty million, perhaps as many as fifty million, Chinese people died. This famine, like many others, was caused by the actions of a human leader. But its horrors were magnified by the release of crop pests from their enemy the sparrow.

Mao's plan failed to account for a phenomenon ecologists call a trophic cascade: changing one species in a food web (say, sparrows) can lead to changes in many other species that it feeds upon or that feed upon it. For ecologists these cascades can be beautiful inasmuch as they obey predictable rules, the invisible laws of nature. Remove a predator, and its prey will thrive. Remove the predator of a predator, and prey become rare.[1] Hans Herren was hoping that the cassava ecosystem would be similarly predictable, that he could find a parasite that would eat the mealybug and so allow the crops to thrive once more. His plan was to do the opposite of what Mao did. He was hoping he could bend ecology's rules in his favor — or, rather, in the favor of millions of Africans.

Efforts at biological control include some of the greatest and most tragic examples of hubris in biology. Beginning in the 1840s, cane toads (*Bufo marinus*) were intentionally introduced throughout the Caribbean to control, initially, rats, then pests that affect sugarcane, particularly white grubs (*Phyllophaga* spp.). The cane toads thrived alongside the rats, which they did not control. By some accounts they did not control the white grubs, either, but they were introduced into Hawaii and the Philippines anyway, where they also might or might not have controlled white grubs and other pests affecting sugarcane. In 1935 hundreds of cane toads were sent from Hawaii to Australia in hopes that the cane toads would control the greyback cane beetle (*Dermolepida albohirtum*), the larvae of which consume sugarcane roots. Only after the toads' introduction did people realize that the animals don't jump high enough to eat the pests. They are, however, doing just fine on other foods and are now devouring rare native species and killing anything that tries to eat them, including rare snakes and domesticated dogs. The toads are ranked by some as one of the greatest threats to Australian biodiversity. We tend to hear more about cases in which biological control agents have unintended consequences than we hear about the far more common cases of failure in which biological control agents have no effect at all.

Advocates of biological control will tell you that these failures are historical—or, as the plant pathologist Harry Evans (see page 95) more strongly put it, "unscientific anecdotes," meaning that the method's success has become far more common than failure. Yet the examples of great successes, successes that changed the world, were, in the 1970s, still few. From the start, cassava did not seem likely to be a success story. For one thing, no one knew anything about the cassava mealybug, not even where it came from. For another, the scale of the problem, geographically, was immense, potentially covering all of tropical Africa and parts of Asia. Yet no other solutions presented themselves.

The first step was to find where the mealybugs had come from. Then, having found that, the next step was to look for animals and other organisms that eat them. But Herren and his colleagues had nothing to go on other than their knowledge about the natural distribution of cassava and the indication from the French biologist Danièle Matile-Ferrero, based on a museum record of a similar mealybug in Brazil, that the pest was probably from the Americas. She could not even pinpoint a continent. This may seem outrageous, that a new and devastating crop pest would emerge out of the unknown. But when most species on earth are unnamed, unnamed species appear on crops and in our backyards all the time.[2] It is ordinary. No one remarked at the unusualness of the situation,[3] only at its difficulty. Herren guessed that the mealybug came from the place where cassava had been domesticated or where the majority of its relatives lived. He just needed to find out where that was.

. . .

In the 1960s, a Texas oilman named Kenneth Lee, along with several geographers, argued that cassava had been farmed at high densities in the Bolivian Amazon and likely in nearby Paraguay and Brazil. Botanists, at around the same time, argued that it was in approximately this same region that cassava had originally been domesticated. These hypotheses together suggested strongly that the first place to look for a strange cassava pest would be in this region. Unfortunately, although it is now clear that they were right, the opinions of these individuals were viewed as heretical by anthropologists who thought that large-scale agriculture could never have been possible in the Amazon. As a result, to the extent that these ideas were present in scientific literature, they were quiet, radical, and obscure. Herren might have nonetheless paid them heed and started looking for cassava mealybugs in Bolivia, Paraguay, or nearby parts of Brazil. He was a bit heretical himself, of course. But he did not.

Herren decided on a five-step plan. He would travel to each of those places where relatives of cassava lived, starting in California. He might have been lucky; the mealybug might have been native to Southern California, and he might have found it right away. It was not. California was a bust. Herren headed south into Mexico, from one patch of cassava to another. He ran afoul of guerrillas and governments in equal measure. In his travels, by his own account, he was arrested, "nearly killed," or both many times. Even if one takes into account the possibility of hyperbole, it seems clear that Herren faced real danger—danger on behalf of Africans he had never met (and their children and grandchildren). When he did, when he was in jungles far from the Switzerland of his childhood, afraid, he must have wondered whether the whole endeavor was lunacy. On farm after farm, in Mexico, Guatemala, Honduras, El Salvador, and Nicaragua, he found nothing.

Then, in northern Colombia, he found something; a cassava mealybug that looked like those he had seen in Africa, eating cassava plants in much the same way as the cassava mealybugs in Africa did. He thought it was the right mealybug, the one plaguing Central Africa. It was not. Nor was this the first time someone had made the same mistake, but even figuring out something so simple as whether it was the right species was a major challenge when so little was known.[4] The bug was new to science. It was yet another new species, another unnamed cassava eater, a species that, whether as insult or praise, bears Herren's name, *Phenacoccus herreni*.[5] The bad news, of course, is that its predators and parasites were likely to be totally ineffective in controlling the cassava mealybug in Africa. They were the demons of a different pest.

Herren went back to searching. He had been to Mexico, Central America, and northern South America. Researchers from the Centre for Agriculture and Biosciences International (CABI) had searched in Trinidad and northeastern South America. The next place to try was Paraguay. To travel to Paraguay, Herren enlisted the collaboration of Tony Bellotti, who was by then working at the Centro Internacional de Agricultura Tropical, in Colombia. Bellotti wrote the book, or at least the major paper, on the pests affecting cassava in 1978 and worked in the years since then throughout the neotropics. Glad to help, Bellotti took Herren to Paraguay with him in the wet season of 1980 to look for the mealybug. They found nothing. Herren begged Bellotti, who would be back later in the year visiting his ex-wife, to check again when it was dry. Bellotti did, and when he did, he found what appeared to be the cassava mealybugs. He contacted Herren. He also quickly sent samples to Doug Williams at the Natural History Museum in London (who had just named the mealybug Herren had found in Colombia). Williams, perhaps the greatest expert on mealybugs in the world at the time, identified the insect as the same *Phenacoccus manihoti* that was terrorizing Africa. They had the right mealybug, thanks in part to Bellotti's meeting with his ex-wife.

More surveys began in Paraguay in earnest in 1981 in a collaboration involving Herren and his team, Bellotti, and researchers from CABI. The cassava mealybug was present in many fields in Paraguay, but nowhere abundant, nowhere destroying whole fields, suggesting that it was being held in check by its own devouring monsters, exactly what Herren had hoped for. Then in 1982, the mealybug was also found in Bolivia, in precisely the region where cassava was not only native but also cultivated at very high densities, high enough to favor the origin of new pests—such as the mealybug and certain pathogens—and perhaps even the species able to control them over the thousands of years during which cassava was farmed there.[6] It was right where it would have been predicted, at least if cassava's history were more generally known.

Bellotti thought he had found the needle in the haystack, the mealybug in its native range. Now he, Herren, and others needed to thread the needle: they needed to find species that ate the mealybug and figure out a way to spread those species across Africa. In Paraguay, they watched the mealybugs in the field. They cut open adults. They poked and prodded. The mealybug seemed to have many enemies, including fungi that devoured its body, ants that dragged it home and fed it to their babies, and, most promising, wasps that laid their eggs inside the mealybugs' bodies. Eighteen species were found preying upon and parasitizing the cassava mealybug.[7] Biologists are biologists because they love life. No organism is good; none is bad. Each is a manifestation of the marvelousness of evolution. Yet for Herren and his colleagues, one life form, the mealybug, was clearly the enemy, and the organisms that consumed the mealybug were beginning to seem a great deal like friends.

Among the most common of the creatures found attacking the mealybug was an animal called Lopez's wasp, *Anagyrus lopezi*.[8] This wasp was promising. It was fecund, specialized (appearing to feed only on cassava mealybugs), and, in its method of killing mealybugs, brutally efficient. Lopez's wasp lays its eggs inside the bodies of the mealybugs. There the larvae hatch and eat the mealybug's blood, then its muscle, then its fat, then its digestive tract, all the while being careful to leave the nervous system of the mealybug intact. The wasp then molts, chews a hole in the exoskeleton of the by-then-empty shell of the mealybug, and flies away to mate.[9]

Scientists at CABI arranged for several Lopez's wasps, along with individuals of seven other parasite and predator species, to be sent to the Natural History Museum in London. From there the parasites were transferred to the CABI quarantine unit, where they were allowed to live through several generations and were minimally tested for their negative effects on organisms other than the cassava mealybug. Having passed these tests, the insects could be taken to Africa.[10]

In 1981, the same year in which the wasps were first found in Paraguay, Herren flew from London to Nigeria with some of these wasps in his luggage, packed in vials stuffed with paper and honey (to keep the wasps alive) and topped with cotton. He got off the plane energized. He'd pop the cap, release the beasts, then watch his winged army rescue a continent. But this was just a daydream; in reality Herren knew that spreading the wasps would require active intervention. He would need to rear many and then spread them himself, or at least spread them with his team, from country to country. This would take time, which those who subsist on cassava did not have. The only substitute for time was money.

Herren thought he would need millions of dollars in order to introduce the wasps at many different places simultaneously. In the event the first batch didn't take, this technique made it more likely that at least some would and that it would happen fast. Herren wanted—needed, he would say—a three-story building, a kind of giant insect fornicarium, in which to breed the wasps. He needed three airplanes to release the wasps once they were ready. He needed assistants to carry out the tedious rearing. Relative to the societal and economic costs of the mealybug, the cost of these needs was minor, but only if Herren were to make his plan work, and no one really believed he would.

He asked for $30 million. He got $250,000 from a UN agency.

Even with this sum, his colleagues thought he had been disproportionately favored in a world where everyone was clamoring for funds
for some good and necessary project. As Herren continue to knock
on doors for more funds, he simultaneously started to move forward, begging forgiveness after each radical action rather than asking permission beforehand. His bosses at the International Institute
of Tropical Agriculture were furious with him half the time and irri-
tated with him the rest of the time. To many, Herren was "another
ecofreak fresh out of Berkeley." The institute wanted to breed new
strains of cassava or spray pesticides rather than deal with biological control. Everyone knew that the odds of success from breeding
new cultivars were much higher than the odds of success from biological control. Everyone also knew that breeding new strains of
cassava would take a long time and that many people would die as
the mealybug spread; this was an inevitable reality of the more certain path. And pesticides were not a realistic option as long as the
goal was to help poor farmers.

Eventually Herren and his colleagues at the IITA raised $6 mil-
lion; it would have to be enough. With this money, they designed
and built rearing facilities twelve feet high, inspired by the hydro-
ponics tower at Disney World. Within these facilities, when everything was working, they could rear one hundred thousand wasps a
week. But most of the time everything did not work; most of the
time, wasps died. And even once the rearing facility was working
there was another challenge: they would need to distribute the wasps
across Africa. They bought a plane. They would use the plane, a
twin-engine Beechcraft once used by the CIA in Asia, to shoot vials
of wasps through the air into farmers' fields. But what would they
shoot the wasps with? They invented a special device, a sprayer that
would propel the wasps out over the fields while alive and intact.
The wasps were the heroes; one had to be careful with their delicate
wings.

Meanwhile, few came to help, and the news crews did not call.
We disregard the slowly building tragedy, the tide rising in slow
motion toward someone else's town. Yet the tragedies that occur
when tropical crops fail affect us all. They affect us as refugees from

hunger flee their native countries. They affect us in terms of our own food supply. You already consume cassava in some processed foods, even if you do not know it. Tapioca is made from cassava, as is most MSG. In other words, even if the plight of hungry families is somehow unmoving, the loss of cassava from mealybug damage is likely to make your pudding and udon noodles more expensive. What's more, cassava is projected to be a crop humans will depend upon more in the future rather than less.

. . .

What will grow in any given patch of dirt on earth is changing. It has been changing for a while. As it changes, the cost of our food will change, and as it does, what we eat will, too. For all but the fabulously wealthy, the price of our food affects what we buy, what we eat, and what gets put into processed foods.

The first change that has affected and will continue to affect what is farmable is exhaustion. The soil in most places we farm has been pushed, pulled, kneaded, and squeezed until all the juice once in it is gone. We attempt to fix this by applying fertilizer, but our abilities are limited. As a result, what you can farm in Minnesota is not the same as it used to be. Corn will not grow in many of the places it grew in the 1950s. We talk about ancient, old-growth forests. One might also describe ancient, old-growth soils. Such soils do not, for the most part, exist anymore, except on steep slopes or other places too difficult to farm. The children of farmers face different choices from those their parents faced.

But compared to what is coming next, exhaustion is a small concern. The bigger challenges are changes in climate. Each cultivar of each domesticated plant species has a specific combination of temperature and precipitation in which it can live. We are changing both temperature and precipitation levels. We can, when there is enough water, modify precipitation. We are rainmakers. But only up to a point, and always with costs. We can't undo the changes to temperature.

As scientists have studied the ways in which plants and animals move, acclimate, and adapt to changes in climate, they have made some observations that seem obvious in retrospect. Species that live at the tops of hills will face particular challenges. The cold climates they need will disappear. Species in the middle ground will move uphill, but into smaller geographic areas than they once inhabited (mountains are narrower on the top). Species at lower elevations may move up. But what happens to the species at the bottom, the species living on the flatlands, where most of us live and where most of our food grows? All things being equal, crops from warm places will move north. Crops currently growing in North Carolina will need to grow in Michigan, and those growing in Michigan may shift to Canada. But what about the vast tropical stretches of the planet? In these places, hot conditions may become hotter and drier. Entirely new climates are predicted to emerge. Under such conditions, we do not know which species will succeed. Nor do we know which species we will be able to farm; such species will have to be tough or die. *We* will have to be tough or die. We will, almost certainly, have to eat more cassava.

Cassava is marginal in more than one way. It is marginal relative to daily Western life, at least for now. It is also marginal in that it grows where little else will. This is what makes it important for the future—more important to you rather than less. No matter whose model of the future you consider, the number of humans who will live with us on earth is increasing. The global demand for food will double by 2050 because of population growth in developing countries—those "marginal" tropics—and the demands of that population. (This doesn't take into account the increase in consumption in developed countries.) These demands will occur in regions that today are generally difficult to farm in. But with global warming, farming in these hot lowland regions will be even more difficult.

Consider Africa. While eastern Africa is predicted to get hotter and perhaps wetter, both northern Africa and southern Africa are predicted to get both hotter and drier. Potatoes are predicted to get harder to farm, as will bananas, plantains, beans, maize, millet, and sorghum. When agricultural scientists such as Andy Jarvis and his colleagues at CIAT, in Colombia, where Tony Bellotti worked, consider the future of tropical agriculture, they predict that there will be more cassava and less of nearly everything else in Africa.[11] Similar realities are likely to emerge elsewhere in the tropics. Cassava grows from nothing, requires little fertilizer, and is robust during dry years and hot years. But all the advantages of cassava would be useless unless Herren or someone else could control the cassava mealybug. All of which is to say that it was not just the fate of those in Africa and Asia that depended on Herren. It was also the fate (or pocketbooks) of all those who buy cassava and its products anywhere in the world or who might buy it in the much warmer future.

Besides Herren and his colleagues at IITA and CABI, the other people attempting to deal with the problem were the farmers themselves. They dealt with it the way that farmers always have: they changed the varieties of crops they were using, favoring the kinds of cassava that seemed best able to withstand the mealybug. Where the diversity of a crop such as cassava is great, this process can be quick. A cultivar that does not die is replanted, traded, sold, and treasured. But in Africa the diversity of cassava varieties was not great. It was a tiny portion of what was present in the Amazon.[12] Yet it seemed to be enough to help. Farmers favored the more resilient cassava, which, as it turns out, tended to be the bitter strains, strains evolved in the Amazon to be toxic to cassava-eating herbivores, including humans.

As West Africans fed more bitter crops to their children, Herren's team tried to breed more wasps, millions of wasps. But Herren himself was coming apart—fraying at the edges from lack of sleep, too much work, and the pressure of believing that a whole continent depended on him—even as everything was coming together. He decided to test the handful of wasps that had already been reared. In November, the dry season, of 1981, Herren released the first wasps in Nigeria near his own workplace. Then he and his colleagues waited. At first nothing. The cassava died. Leaves wilted. Farmers cursed both the mealybug and Herren and his team.

Then the news seemed to change. Some of the fields were doing better; some of the mealybug populations were doing worse. The wasps were a success in the farms in which they were released in Nigeria. The cassava grew strong, healthy, and green. In addition, the wasps survived the wet season. They made it to a second year. Wasp releases were done again 1982. Those wasps, too, would survive the rains.[13]

Amazingly, the next steps were also successful. The wasps bred well in captivity. They could be released on the ground or, rarely, from the plane. Soon Herren and his colleagues released them both on the ground and from the plane in nine countries. By 1986, wasps were released in fifty sites, a number that would grow to 150.[14] Some countries begged for wasps; others quietly permitted them. Cameroon declined but got wasps anyway when they flew in from neighboring countries.[15]

In the meantime, the mealybug plague had grown far worse and more geographically expansive in places where the wasps had not been released. By 1983, cassava mealybugs had reached the status of an outbreak. In Ghana, farmers lost 65 percent of their yield (a $58–$106 million loss at the time). The price of cassava in the market increased ninefold. The price of planting material—new cassava—increased by a factor of 5.5. Where the mealybugs were present, farmers had to plant two or three times as much crop in order to get the same amount of food as they harvested the previous year. Often they did not have the land, or the time, to do this. Hunger loomed; to those who remembered history, events bore a striking resemblance to the buildup to the great potato famine, except that rather than just an island a whole continent was at stake. A lot depended on the tiny wasps.

Finally a new kind of news started to come in, news of mealybugs disappearing in places far from the wasp release sites. At last the wasps were spreading! By March of 1983, just two years after the first release of wasps (ten wasp generations), wasps could be found in every cassava field within a hundred kilometers (sixty-two miles) of each of the release points. Not only was this fast, it was also the fastest rate of spread ever recorded for a parasitic wasp of any kind.[16]

Millions of released wasps turned to billions in a few years. Cassava grew back, almost magically. In Nigeria, cassava production increased from fifteen million tons before the wasps to forty million after them.[17] Across Africa in general the change was similar. A tiny wasp had saved millions of people, potentially hundreds of millions, thanks to the man who midwifed them across continents and the relatively few others on which so much had come to depend. By conservative estimates, the benefits of the biological control program relative to its cost could be expressed as a ratio of 149 to 1.[18] Less conservative estimates put the benefit-cost ratio at 1,592 to 1. And these estimates don't even include the added benefit of not having to use pesticides.[19]

There is no caveat to this story; it is just an unadulterated success. Cassava mealybugs still exist in African cassava, but they are kept in check by the trophic level above them. This is not the balance of nature. Nature is far more often in open war than in balance. It is instead something far more precious and yet attainable: the balance of agriculture.

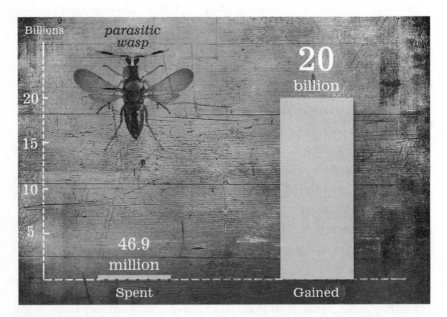

Figure 5. Dollars spent introducing parasitoid wasps to Africa to control cassava mealybugs compared with dollars gained by African economies because of the introduction. Estimates are drawn from Jürgen Zeddies, et al., "Economics of Biological Control of Cassava Mealybug in Africa," *Agricultural Economics* 24, no. 2 (January 2001): 209–19. *Figure by Neil McCoy, Rob Dunn Lab.*

. . .

Thanks to Herren and the more than three hundred people who worked with him on the mealybug project, a pest that threatened to produce an African famine of unheralded proportions was controlled. By 1987, the wasp had spread to virtually every farm growing cassava in West Africa. All this happened in a little more than a decade. One can debate why the late blight killed so many in Ireland, why the potato famine was so boundlessly tragic. With the story of the cassava mealybug there is no such ambivalence. Millions were saved because of the work of hippie biologists who believed in biological control, a bunch of entomologists, an ex-wife, and a previously unknown wasp species that continues to do its biocontrol work on our behalf to this day.

Meanwhile, in 2008, the mealybug was found for the first time in Asia—in Thailand, where it had likely spread via the cuttings of cassava stems.[20] By 2009 it had spread across seven hundred square miles. By the time it was discovered, Tony Bellotti had retired (he has since died). Herren had moved on to leadership roles (though he continues to work). Danièle Matile-Ferrero, the taxonomist who gave a scientific name and formal identity to the cassava mealybug, had retired (and hence was no longer being paid), but she was still doing her work.[21] What one might hope is that, in their places, one could find a new generation of experts poised to play similar roles, along with even more support, more funding, and better facilities. The good news is that a new generation has been trained, a generation on which we might hope to depend. Yet when Thai scientists found the cassava mealybug, they brought in one of the retired superstars of the earlier generation, Bellotti, to help. The Thai scientists, in consultation with Bellotti, developed large-scale rearing facilities for Lopez's wasp. They released the wasps. The wasps have spread successfully.

In this story of the cassava mealybug in Thailand, one finds another round of heroic work by unheralded biologists. But one also encounters a reason for humility. The cassava mealybug is a known problem yet still could not be kept from arriving in Thailand (or, subsequently, Vietnam, Malaysia, Indonesia, and Laos). Nor should we forget the other humbling aspects of the story: when the cassava mealybug was found, it was not yet named; in looking for the cassava mealybug, a handful of other new cassava mealybug species were found; and among the pests attacking those species—the secret agents of biological control—virtually none of them had names, either. In an ideal world, if we want to keep our crops healthy, we should have full lists of the pests attacking our crops and great zoos of the organisms that can be used to control them, zoos at the ready when they are needed. We don't. Cassava may well be the future of tropical agriculture. If it is, a relatively small group of entomologists has played an outsize role in allowing such a future to be possible.

Meanwhile, we should ready ourselves for the next wave of pests. In the Democratic Republic of the Congo alone, according to Sarah Gurr, more than 95 percent of the pests and pathogens that can (and likely will) arrive from their native lands have yet to appear. And Gurr studied only those pests and pathogens that have already been discovered and named. In the days before the cassava mealybug was noticed in West Africa, a study like Gurr's would not have noted it as one of those yet to arrive. It is too hard, and it costs too many lives, to try to understand the pests that attack our crops only during emergencies. We need to do thorough studies of these pests in advance. The monsters need to be named and known. How can we go about this?

We need to survey all the traditional crops we can, particularly those not being sprayed with pesticides, and find the pests on them. We need to survey those pests and find the species that eat them. While we're at it, we need to understand the mutualists of these crops, their partners—the animals that pollinate their flowers (the pollinator of cassava remains unknown), the fungi that live in and defend their leaves, the microbes that help their roots extract nutrients from the soil. Some of these studies may require us to go to remote communities. Or, alternatively, we can engage families in those communities to help us to study the species associated with traditional crops. This may seem like a difficult proposition, but think about the case of a crop native to North America, squash. Squash is pollinated by a specialist bee that depends on the squash plants and has, it's been shown, spread as squash plants have been moved.[22] But the pests affecting squash and their enemies are less well understood and in some cases haven't even been named yet. It would not be hard to distribute seeds to kids in schools across the United States so that they could plant traditional varieties of squash and, in doing so, document pollinators, pests, and natural enemies. It just hasn't been done yet.

Once we know the species associated with each crop, we need to figure out how to grow them, study them, understand them, and use them to our benefit. How many species might this include? Gurr considered several thousand pests and pathogens but was quick to acknowledge that many more remain to be discovered and named. Let's say, conservatively, that four thousand species of pests and pathogens exist and that each of those pests and pathogens falls victim to five parasitoids or pathogens capable of controlling it. Then we are talking about twenty thousand species of pests, pathogens, and enemies that need to be studied, saved, cultured, understood, and held at the ready for the time when they are needed. This is to say nothing of the mutualists; crop mutualists are an even greater mystery and might include hundreds of thousands of species.

Many grand challenges in biology are not achievable. This one is, yet we are nowhere close to achieving it. We have barely begun to try. This is not a critique of those who work in the field identifying these species and figuring out their biology, their needs and limits. It is instead a critique of what we have chosen to invest in as a society. It is a critique that is necessary, because while the incremental benefit of studying some insect or fungus species might be small and slow to accrue, the loss of not having studied it can be enormous and immediate, as has proved to be the case with cacao, one of the most endangered crops on earth.

6

Chocolate Terrorism

Scientists used to study the world to understand it; we study it in order to change it.

—Karl Marx

In the moments before the collapse of an industry or a civilization, we tend to assume there are warning signs, omens, and signals. But when a pathogen or pest strikes, the moment before the fall might look like any other day. It might look like May 22, 1989, in Bahia, Brazil, on the Conjunto Santana cacao plantation. The rain-forest birds were singing, loudly. The trees were dripping with rain. Fat cacao pods hung from the cacao trees. The air smelled of sweet rot and the richness of life. Nothing was wrong. Everything was alive. Then one of the technicians on the plantation saw something unusual, a cacao tree branch swollen with some kind of cancer.

The Conjunto Santana farm was one of the largest cacao plantations in Brazil. That also made it one of the biggest in the world. It was owned by Francisco Lima, a.k.a. Chico Lima. Lima employed many men and women to pick and open the fruit, ferment the seeds, and take each of the other initial steps necessary to make cacao into an urban delicacy. In the process, Lima became relatively wealthy and powerful. His power extended beyond the cacao plantations and into politics. He was elected a local leader in the UDR (União Democrática Ruralista, or the Rural Democratic Union), a political party that sought to protect private property such as the plantations owned by Lima and the other so-called colonels of cacao. Lima was in control of his life. The discovery of the swollen branch was the first clue that he might be about to lose control. He was most worried that the cancer was a particular pathogen called witches'-broom, which, he knew, could destroy whole plantations.

Lima had the cancer sent to CEPLAC (Comissão Executiva do Plano da Lavoura Cacaueira, or the Executive Committee for the Cocoa Farming Plan), the federal organization in charge of helping farmers of the chocolate tree (*Theobroma cacao*), for identification.[1] CEPLAC was founded in the late 1950s in response to a crash in the cacao market, initially as a way to devise a financial support plan for the landowners who were in debt.[2] As CEPLAC expanded, so did its aims: to modernize production—more trees per acre, more pesticides, more fertilizers; to plant more productive (and likely less resistant) varieties of cacao; to conduct research; and to monitor cacao plantations. It would ultimately become a cacao research, extension, and credit agency all in one (at least until the late 1980's). CEPLAC helped cacao rebound and then increase, both in terms of total production and in terms of the area covered with cacao trees.[3] As a result, by 1980, cacao was one of the most important exports not only of Bahia but also of Brazil in general. If anyone could help, CEPLAC could. On May 30, eight days later, CEPLAC called Lima. The cancer was indeed witches'-broom—or, as it is known in Portuguese, *vassoura-de-bruxa*.

Witches'-broom is a disease associated with a fungus (*Moniliophthora perniciosa*).[4] It invades trees through wounds and stomata (the puckered holes on leaves through which trees "breathe"). Once inside, it begins to consume tree tissue and alter tree growth, causing the tree to grow grotesque broom-shaped tumors (the cancer), hence the name of the disease. With time and the passage of seasons, the trees with brooms succumb to the infection. When they die the cancerous brooms blacken and produce pink mushrooms from which millions of spores are released into the air, out toward other trees.

By the time witches'-broom reached Lima's farm, it had already proved its ability to spread rapidly. Witches'-broom is native to the western Amazon—Peru, Bolivia, Brazil, and Ecuador. In the late

1800s it was accidentally introduced in Suriname, leading to severe losses in cacao production there. By the 1920s it had been accidentally introduced in Trinidad. Once it arrived, the export of cacao from Trinidad declined from thirty thousand metric tons per year to eight thousand tons per year in less than a decade. Then there was the case of Ecuador. During the late 1800s and early 1900s Ecuador was the king of cacao production both in terms of quantity and perceived quality. Ecuador's cacao was special, based on a wild variety of cacao, Arriba (also called Nacional). It was planted beneath the canopies of wet rain forests to the west of the Andes (rather than to the east of the Andes, in the Amazon, where witches'-broom was present). In small towns, it is said that owners of large Ecuadorian cacao haciendas became affluent enough to send their clothes to Europe to be washed.[5] Then the pathogen that causes frosty pod rot, *Moniliophthora roreri*, was introduced to western Ecuador, likely from Colombia. Once frosty pod arrived, witches'-broom came shortly thereafter, introduced from the eastern Amazon. The consequences were swift and catastrophic for the Ecuadorian cacao industry. The production of cacao in Ecuador went from forty thousand metric tons per year to fifteen thousand. Nearly all of the large cacao plantations were abandoned, and the average yield per hectare of those smaller plantations that remained decreased from one thousand kilos to less than three hundred.

But neither frosty pod nor witches'-broom had arrived in distant and isolated Bahia. Bahia is in the Atlantic Forest region of eastern Brazil. This region shares a climate with parts of the Amazon, where witches'-broom is native, but the two regions are separated by dry, scrubby shrublands. For rain-forest species such as cacao (and witches'-broom), these shrublands, which span a distance nearly as great as that between New York City and Los Angeles, are as inhospitable as a sea, a barrier to both survival and movement. It was a sufficiently harsh barrier that even if one region but not the other had a particular pathogen, such as witches'-broom, it could well stay that way. Such had been the case for millions of years. Then came that moment on Lima's farm in May of 1989.

. . .

Cacao was first planted in Bahia in 1746 by Louis Frédéric War-neaux, a French merchant.[6] Warneaux is said to have brought the seeds from Pará, in the Amazon, to Bahia in his pocket (pockets feature prominently in the history of seeds and their movements). Each generation, their numbers increased, especially once the demand for chocolate grew in Europe and North America in the late 1800s in response to increases in wealth associated with European industrialization. Between 1840 and 1890, exports of cacao from Brazil increased by nearly 300 percent, almost all of it from Bahia.

In Bahia, Warneaux's trees and their descendants were blessed with a favorable climate, but they also had the benefits of isolation and a reprieve from their pathogens and pests. This release from enemies was unusual in that it occurred not on some distant conti-nent but rather in an isolated part of the continent to which cacao was native. Brazilians could grow a crop in the Americas with a yield as great as might be achieved in, say, West Africa or tropical Asia. That *Theobroma cacao* never made it on its own, with all its pathogens and pests, to Bahia was a function of the isolation of the region.[7] The specific variety of cacao grown in Bahia was not chosen for its flavor, its perfect suitability to the region's soil, or its resis-tance to pathogens and pests. It was simply the variety that War-neaux happened to bring over, a variety of Criollo cacao called Amelonado. It also happened to be productive. As with potatoes, cassava, bananas, and so many other crops, cacao's diversity was low.[8]

Because the Amelonado variety of cacao Warneaux carried over grew so well, little time or money was invested in finding better varieties. Agricultural homogeneity was an advantage as long as pests and pathogens could be held at bay. It is for this reason that CEPLAC focused less on diversifying cacao varieties than on strict control of the movement of cacao. Similarly, farmers planted the highest-yielding variety, rather than the diversity of varieties that might best shield them from future risks, including witches' broom.

The isolation of Bahia's cacao from witches'-broom and frosty pod rot was the secret to its success, to Bahia's success.

In theory—and, Lima hoped, in practice—witches'-broom might be controlled, maybe even eradicated completely, before it spread on his farm or to other farms. Gone were the days when one had to make a case that fungus was to blame. Now fungi are readily recognized as culprits. Fortunately many fungi can be controlled with fungicides, but those agents are more useful as preventives than they are as treatments. Once a fungus has grown into a plant, it is hard to kill it without killing the plant.[9] The plantations to the west, in the Amazon, were for the most part too poor to invest in fungicides, much less devise innovate ways to use them. In the Amazon, witches'-broom and the annual loss of 30 to 60 percent of the bean production of cacao trees were treated as facts of life, the ordinary hardships of getting by. But inasmuch as Lima and others like him in Bahia had more money than cacao farmers elsewhere, they might have more options for dealing with the problem.

Yet a scientific paper published around the same time that witches'-broom was found on Lima's farm offered worrisome insights. The use of fungicide was difficult even as a preventive form of control because it washed off in the near daily rain and couldn't readily get to the fungus in the tree. Also, the fungus produced spores and promptly recolonized any treated tree. The best available fungicide was the same copper sulfate used to treat potato blight.[10] But even copper sulfate was only modestly effective.[11] CEPLAC was going to have to do something on Lima's farm that no one had done before on a large scale.[12]

The fate of his farm was uppermost in Lima's mind. The potential consequences of the spread of witches'-broom to other farms in Bahia, however, were far larger. In 1989, Brazil was the second-biggest cacao producer in the world, vying only with Ivory Coast. Cacao was the country's second-most-important export commodity. The loss of cacao plantations would affect the country's economy, the supply of cacao, and the lives of the plantation owners and their workers. It would also affect the region's biodiversity, a biodiversity that traditional cacao plantations helped sustain.

. . .

Many aspects of farming cacao are similar to those of farming other tropical crops, but in one way farming cacao is quite different, more similar to farming coffee than anything else. Both coffee and cacao are understory trees. In the natural forests in which these trees are native, cacao and coffee never grow tall enough to get their branches and leaves above those of other trees. While some rain-forest trees grow to more than seventy meters (230 feet) in height, cacao and coffee seldom reach more than ten meters (thirty-three feet). These trees evolved to take advantage of the chance flecks of light among the shadows. Living in the understory is difficult, and so understory trees tend to depend disproportionately on other species for help. They depend on webs of interactions with other species that are easily disrupted and that we are only beginning to understand.

Cacao flowers, like those of most tree crops—including coffee as well as pears, peaches, and apples—depend on insects for pollination, to carry the pollen from the male parts of one flower to the female parts of the same flower or another. Cacao roots, like those of most tropical trees, depend on partnerships with fungi to reach each and every soil nutrient available, even those seemingly too far away or in cracks seemingly too narrow to probe. Cacao leaves depend on fungi called endophytes (where *endo* means "inside" and *phyte* means "plant"), which grow inside their tissue to defend them from pathogens. Finally, cacao seeds depend on animals to carry them away from their mothers to places less plagued by pathogens and pests.[13] In order to get away, cacao invests in large fruits that are carried by monkeys (whereas coffee trees produce bright red berries that are ingested by birds).

Cacao's success is embedded in a web of interactions with other species, many of them poorly studied. As a consequence, the effort to control witches'-broom and other diseases depends not only on the biology of witches'-broom itself but also on the biology of many or even all the other species with which cacao interacts. What's more, these species, in the years before witches'-broom arrived, had started to change as CEPLAC began to favor large-scale, intensive approaches to farming. This change meant that not only was the cacao living in different conditions from those it might have experienced historically but all the other species were as well. The ancient system had been destabilized. Add to this one more complexity: although witches'-broom had been discovered years before, it was not yet well understood.[14]

. . .

Fortunately, in Bahia, the witches'-broom, regardless of the details of its biology and horrors, was still found on just one individual tree. On May 26, 1989, only four days after the pathogen was detected, CEPLAC set out to wall off the problem, to create a bubble around Lima's farm out of which the witches'-broom could not pass. They started with a survey to find out how bad things were. Thirty agricultural workers from CEPLAC walked the farm, checking each of the nearly one hundred thousand individual trees—miles of cacao trees beneath miles of forest. One by one they found bad news. At least 112 trees were visibly infected—that is, they had produced mushrooms—almost certainly suggesting an even larger number of invisibly infected trees. One simply had to hope that whatever the number of invisible infections was, it was small and geographically circumscribed by the area in which the infected trees were found—the hotspot of contagion. The infected trees were sprayed with a fungicide, killed with an herbicide, removed with a bulldozer, and then burned. A buffer of trees was cut around the area in which they had been found, and the other trees in the vicinity were monitored weekly.[15]

As weeks passed, the bubble around Lima's farm remained unbroken. There seemed reason to hope. Then twenty-one infected trees were discovered outside the quarantine area on Lima's farm. In response, CEPLAC made a firm statement. The plantations where the witches'-broom was found must all be cut down and burned in their entirety to prevent contamination. The rest of Lima's farm was felled—each cacao tree and then the larger trees that shaded them. He had his land cleared for the common good. He had no choice.

CEPLAC implied or even said (depending upon who is telling the story) that if he did not allow his trees to be cut down he would be held criminally responsible. Clearing Lima's plantation would require more than fourteen thousand man-hours of work between May and November of 1989. Ninety-eight thousand cacao trees and more than ten thousand rain-forest shade trees were destroyed. As the end of October neared, CEPLAC scientists felt they had contained the witches'-broom. In the buffer area around the denuded landscape where the trees had been destroyed, no new infestations had been detected in the 120,000 trees laboriously checked by hand. Then came more news.

On October 26, a farmer at the other end of the cacao region of Bahia, one hundred kilometers (sixty-two miles) from Lima's farm, also discovered witches'-broom. Like Lima, he was one of the farmers with a large amount of land planted with cacao and so had much to lose. This second case was worse. The disease appeared to have been present for months, perhaps years. Soon a handful of other cases were detected, and these new cases were even more unusual.

On one farm, a tree infected with witches'-broom also bore a branch from another cacao tree, even more covered with witches'-broom mushrooms than the host tree. Some said it had been tied to the tree with a piece of knotted rope.[16] Given enough time, nature can do many things. But nature does not tie a knot. In the coming weeks, other branches were reported on other farms, tied, it was later said, with similar ropes.

In the months to follow, more infected trees appeared on more farms. CEPLAC took more desperate measures. Everyone, they said, must cut down all their cacao trees when infected trees were found on their plantations. Everyone must try to graft a new variety of cacao suggested by CEPLAC. Loans would be available, contingent on each of these actions. But none of that really helped. As a result, the cacao production of the region, then the cacao trees, then trees in general began to disappear. What followed was the kind of devastation seen in other landscapes after pathogens arrive.

By 1991, witches'-broom had spread throughout the cacao-growing region, along Highway BR-101, along smaller roads, along

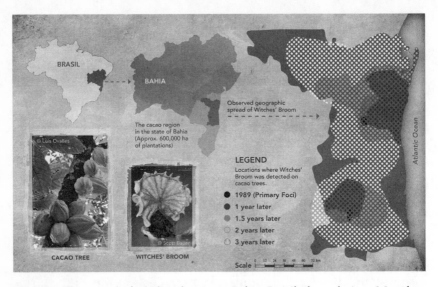

Figure 6. The spread of witches'-broom in Bahia, Brazil, through time. Map data derived from J. L. Pereira, L. C. C. De Almeida, and S. M. Santos, "Witches' Broom Disease of Cocoa in Bahia: Attempts at Eradication and Containment," *Crop Protection* 15, no. 8 (December 1996): 743–52. *Figure by Neil McCoy, Rob Dunn Lab.*

footpaths, everywhere. The acreage planted in cacao in Uruçuca, the region of Lima's farm, declined by 75 percent. A similar fate befell other regions.[17]

. . .

Witches'-broom could have been brought accidentally from the Amazon to Bahia on the sole of a boot or in a sheathed machete. Many cacao farmers had plantations in both places—healthy, productive cacao plantations in Bahia and less productive and more afflicted plantations in the Amazon; accidental transportation from the afflicted to the healthy farms would be easy (as the case of the cassava mealybug makes clear). Yet the knots, if the stories of their presence were true, suggested the agency of humans. They suggested intentionality, a kind of agricultural terrorism.[18]

In general, crops worldwide are poorly protected, poorly monitored, and geographically expansive.[19] The same features that make agricultural terrorism possible—easy, even—make it hard to detect.

Someone, it seemed, had introduced witches'-broom to the cacao plantations of Brazil. But if so, who? One possible candidate was someone from Ivory Coast, Ghana, or Malaysia, where most of the cacao not grown in Brazil was grown—someone bent on destroying Brazil's cacao production in order to benefit his or her own country. Although the possibility was far-fetched (especially since a West African wandering among the cacao plantations would be conspicuous), it was not impossible.

Whereas few countries have been caught carrying out agroterrorism, many countries have readied themselves to attempt it. But unless a country or an individual confesses to agroterrorism, confirming cases is almost impossible. During World War I, the German military devised methods of introducing pathogens into grain supplies that would spoil the grain of enemies. During World War II the French discussed releasing both Colorado potato beetles (*Leptinotarsa decemlineata*) and late blight from airplanes onto German potato fields. This program ended when the Poudrerie Nationale laboratories in Le Bouchet, France, where these projects were being readied, were taken over by the Germans. The Germans then took up where the French left off. In 1943 the Germans released live potato beetles and wooden models of potato beetles from planes as a test of how far the beetles might spread (they did this within Germany, amazingly). Mass breeding of the beetles also seems to have begun.

Before, during, and since the war, the United States has worked on biological agents of agricultural terrorism for use against each of twenty or thirty different crops. Among them were two pathogens affecting rice—rice blast (*Pyricularia oryzae*) and brown spot (*Cochliobolus miyabaenus*), both of which appear to have been tested in aerial releases. Japan, in turn, along with biological weapons against humans, also developed many biological weapons against crops, though these seem to have not been used. The Soviets had a large program to produce crop pathogens for use in war. Though the details of this program have yet to be revealed, it, too, included rice blast. It also included stem rust. After World War II, most countries stopped their agricultural weapons programs, except the United States and the Soviet Union. The United States studied ways to release stem rusts of wheat and rye (*Puccinia* f. sp. *secalis*), rice blast, and late blight. The stem rusts were developed to the point of being a fully ready weapon, to be delivered in bombs filled with turkey feathers coated in spores of the rust. In order to make the weapons, to be dropped in Korea during the Korean War, spores were produced in huge quantities. Eventually rice blast was also made ready for release. This work continued until Richard Nixon renounced the program and destroyed its stockpiles through an executive order in 1973. The more secret but even larger-scale Soviet program was not discontinued until 1980. A decade later, no weapons of mass destruction were ever found in Iraq, but weapons of agricultural destruction were. The Iraqis were developing two pathogens that attack wheat (*Tilletia tritici* and *T. laevis*), both of which were ready to be released. Could the witches'-broom in Bahia be a case of successful agroterrorism, perpetrated by one or another West African country seeking to gain at the expense of Brazil?

Ivory Coast and Ghana had the most to gain financially from such an act. Yet they were unlikely culprits. Whoever planted the witches'-broom fungus on the trees had a sophisticated understanding of both local geography and local plantations. The sticks were tied on trees in places (along rivers and wind courses) where the dispersal of the sensitive spores would be maximized. Witches'-broom spores also need water to survive. Someone appeared to know these details. In addition, cacao farms in Ghana and Ivory Coast, in contrast to those in Bahia, are run by many thousands of small farmers. Even if Ghana or Ivory Coast might, as a whole, benefit from the demise of Brazilian cacao, the same was not necessarily true of an individual farmer.

Also, with the advantage of hindsight, one notes other oddities. Interviews with CEPLAC workers during the filming of a short documentary about the crisis suggest that a bag of infected cacao branches was left in the headquarters of CEPLAC with a note said to have read, in Portuguese, "Go look for the rest of the witches'-broom, you bums." The note, those interviewed claimed, was placed on a desk inside CEPLAC's headquarters, a location one could only reach after passing through security. Who was the note for? Who had written it? Was it a threat or something less obvious?[20]

Another strange feature of this disaster was that some of the steps CEPLAC told farmers to take seemed to actually make it harder for farmers to recover. Likely this represented nothing more than the challenges of dealing with a new and difficult pathogen along with the ordinary incompetence of any institution in light of a disaster, but desperation made it easy to see a conspiracy in each failure.

Then there was the issue of replanting. Farmers who destroyed their trees were told to use only the CEPLAC trees to replant, but the CEPLAC trees were all clones, one of five genetically identical lines. Those clones proved of modest resistance to witches'-broom. And, they were genetically identical; they needed to be planted near other clones in order for pollination to occur. Many cacao varieties, like many plants in general, only produce fruit if their flowers receive pollen from genetically different individuals. This is a means of preventing inbreeding where it isn't otherwise obvious who one's cousins are. CEPLAC had designated two of the clones as good pollinators and those were to be planted with the most resistant clones to assure both at least some resistance and pollination. Unfortunately, because farmers were focused on their yield—more pods, more seeds—they preferred to plant the clone with the largest pods and largest number of seeds at the expense of planting many (or any) "pollinator" clones. As a result, little pollination occurred in the newly planted cacao plantations, but this only became clear several years later, once the clones were old enough to flower—and after a great deal of tending and work and money had been invested. Finally, CEPLAC did not help the farmers financially with any of this, but instead mediated big bank loans at high interest rates, which would ultimately prove very difficult to pay back. The farmers understood they were taking the loans, and did not always choose wisely in terms of what those loans were spent on. Nonetheless, when cacao failed to recover fully the inability of the farmers to repay the loans and interest made those who offered the loans seem to be easy and unambiguous villains.

The net result of it all, the mix of the bad pathogen, bad luck, bad decisions, and difficult economic conditions and incentives, was collapse. In 1989, Bahia had six hundred and fifty thousand hectares of land in cacao production. By 1992, just four hundred thousand hectares remained, and none of those parcels of dirt and tree was producing as many pods of chocolate as it once had. Production declined by 75 percent and then, because there is still no real reprieve from witches'-broom, never recovered.

If the consequences of witches'-broom were just biological, recovery might have occurred. Varieties of cacao that are tolerant (if not fully resistant) to witches'-broom have now been bred, and CEPLAC has developed approaches to mitigating the pathogen's effects and spread. Unfortunately, the effects were not just biological. As farms collapsed, the consequences cascaded. Two hundred thousand workers lost their jobs and, with their jobs, their support systems. The cacao plantations had provided housing and, in some cases, schools; now both were scarce. Farmers and farm workers are said to have committed suicide—with their guns, by hanging, with rat poison. One farmer, distraught, just walked around town for days, crying.[21] Many tens of thousands of workers immigrated to bigger cities in the region. These cities, by then full of unemployed and hard-to-employ farm workers, saw increases in crime and drug use. Having cut down their own cacao plantations, the barons felled additional forests so that they could sell the wood and make a little money. Many rare species became even more rare.

The Brazilian empire of cacao fell more quickly and completely than anyone seemed able to imagine.[22] When the first infected tree was discovered, Brazil was the number-two chocolate-producing country in the world. Just four years later, it was a net importer of chocolate (as it remains). The federal police investigated the case and soon abandoned it. Were the police themselves involved? The more people thought about it, the more likely a local culprit seemed. But who was it, and why wasn't there more of an investigation? If this really was agroterrorism, the intentional destruction of a crop, it would be the worst case of such terrorism in history. Yet there it sat, unresolved, and most assumed it would remain so forever. They were wrong.

. . .

The surprise answer came more than a decade later. In 2006, a man no one connected to the cacao crisis in Bahia had ever heard of came forward with an announcement. In one of four interviews with the journalist Policarpo Júnior in the popular magazine *Veja*, he said, "My name is Luiz Henrique Franco Timóteo. I was one of those responsible for the introduction of witches'-broom in Bahia." He destroyed, he would go on to admit, the cacao of Bahia, Brazil, and with it so very many lives—and he had done it on purpose. What's more, he said, he carried out this act of agricultural terrorism easily.

Timóteo claimed that in 1987, two years prior to the first witches'-broom infection, he was in Cacuá, a bar in Itabuna, drinking.[23] There, he said, he met up with five other men—Geraldo Simões, Wellington Duarte, Eliezer Corréa, Everaldo Anunciação, and Jonas Nascimento—all of them technicians at CEPLAC. The men were members of the left-leaning populist Workers' Party (Partido dos Trabalhadores, or PT), which was in favor of the redistribution of land and wealth, a party opposed to the power of cacao barons. They intended to undermine the barons' economic and political clout by destroying the cacao industry.[24]

The six men had, Timóteo said, clear motives for their actions. They were not owners of cacao plantations and so were excluded from power in the region. The only way to achieve more power was to weaken the influence of cacao. The easiest way to weaken cacao's influence was to kill the plant itself. These men, in Timóteo's telling, felt themselves to be revolutionaries bringing the power back to the people. They were not destroying the industry of a region; they were seizing the economic power of the owners for the people. This effort even had a name: Cruzeiro do Sul, or Operation Southern Cross.

Gathered together, they would have been aware of the storyline they were enacting. It was a story out of a novel, a specific novel. Published in 1958, Jorge Amado's *Gabriela, Clove and Cinnamon* is set in Ilhéus, Brazil, in Bahia, in the 1920s. In it, the old cacao barons run everything, including local government and the law.

They pay or own everyone. Then a new guy comes to town with aims of turning this power structure on its head.

Of the six men, Timóteo knew the Amazon best, so he was to be the one to travel there and gather witches'-broom to be used in the plan.[25] He rode in a bus more than fifty hours to Porto Velho, in Rondônia, Brazil, where he was born. Once there, he gathered branches infected with witches'-broom and hid them in rice bags for the trip home. He made it past the quarantine checkpoints with ease, not just once but several times, over several trips, during which he transported a total of around 250–300 infected branches.

Timóteo claimed that once back in Bahia he and the five other men drove up and down the region's major north-south road (BR-101) in a car with the CEPLAC logo on it (to reduce suspicion if they were detected, though the truth was no one paid attention to them at all), tying the sticks, with delicate little knots, to trees along the way. They targeted, disproportionately, the farms of the two most powerful growers, that of Lima in Uruçuca, and the other owned by Luciano Santana in Camacan. Lima was targeted not only because his plantation was large but also because of his political role as the head of the Rural Democratic Union, powerfully opposed to land reform and in favor of presidential candidates who were opposed to it, too. It was, of course, on Lima's plantation that the first infected tree was detected. And the second was on Santana's plantation.

If Timóteo's story was true, the loss of an entire nation's cacao was linked to the actions of a tiny handful of men. In Timóteo's telling, members of CEPLAC, the organization whose goal was to help with the cacao, were in on the sabotage, bringing into question the actions CEPLAC took to control the spread of witches'-broom. CEPLAC's first action was to burn the trees belonging to Lima "for protection," but in Timóteo's telling these fires were vindictive, meant to stop the success of Lima rather than the witches'-broom. This seems unlikely. Those consulted to develop the eradication plan were plant pathologists from not only CEPLAC but also from elsewhere in Brazil (including the renowned plant pathologist João Luis Marcelino Pereira) as well as Colombia and Ecuador. As a result, the simpler, and in many ways more ominous, conclusion is that the failure of control was due not to maliciousness, but instead the very challenges of stopping a newly arrived and rapidly spreading pathogen even when the right experts have been gathered together.

As for those whom Timóteo implicated as his fellow saboteurs (all of whom have denied involvement), the social change precipitated by the demise of the cacao economy heralded their success. Geraldo Simões was elected mayor of Itabuna and then secretary of agriculture for all of Bahia. Everaldo Anunciação would become deputy director of CEPLAC. Wellington Duarte became superintendent of Bahia and later deputy director of CEPLAC, Eliezer Correa became chief of CEPLAC's programming and planning, and Jonas Nascimento was appointed adviser to the head of CEPLAC's Extension Service. But to the extent that Timóteo cared about the public for which he and those he implicated fought, the act was a tragedy. A quarter of a million people lost their jobs. Nearly a million people, including the employees of the plantations and their families, moved to the cities. Perhaps it was with these people in mind that Timóteo finally confessed. He felt, he said, great remorse for what he had done. Among those affected by his actions were his own relatives, people he loved whose lives were crushed. His cousin, a man who was bringing in thirty million reals a year (roughly eight million dollars), lost everything. That same cousin heard Timóteo's confession. The cousin told Timóteo to disappear, warning that people would come for him.

Timóteo and the five he implicated have gone on with their lives. The remaining cacao growers, those with enough power or money to still have a voice, requested that Timóteo be tried. In 2007 a six-month trial ensued in which it was concluded that yes, definitely, there had been criminal introduction of witches'-broom and that yes, Timóteo had been involved. It was also concluded that there were "irregularities committed by the employees of CEPLAC," including attempts to hide or fabricate documents relevant to the case. But no conclusion could be drawn, the jury decided, as to the guilt of Timóteo's alleged coconspirators without more evidence. Also, the statute of limitations had lapsed. The statute of limitations on biological terrorism in Brazil is eight years.[26]

The rest of the story is ambiguous, left in the fuzzy realm of jungle tragedies. Subsequent genetic studies have shown that the two strains of the fungus in Bahia do, as one would suspect if Timóteo were telling the truth, derive from the eastern Amazon, near the border with Bolivia. Other reports written since Timóteo's announcement tend to give credence to his role in the introduction. It is hard to imagine a reason for falsifying such a terrible admission. Timóteo's guilt is important to assess, if only to better understand one of the few cases of agricultural terrorism in which everyone agrees that a crime has been committed.

Since the act to which Timóteo has confessed, little about our ability to deal with witches'-broom has changed. The best strategy has been to search for varieties of cacao that are more tolerant of the fungus. Thirty percent of the cacao in Bahia is now of a more resistant variety. No varieties that are fully resistant have yet been found. Fungicide helps, but it is expensive.

Another approach has, however, emerged, the result of comparing the biology of cacao in the wild in Pará, Brazil, to that on plantations. This comparison revealed a fungus, *Trichoderma stromaticum*, that attacks witches'-broom. The fungus was found growing on witches'-broom fungus (which, in turn, was growing on cacao). In the brooms and pods, it infects the witches'-broom, releases enzymes that digest witches'-broom, and, in doing so, prevents witches'-broom from forming mushrooms and spreading its spores. CEPLAC is now growing this fungus and selling it to farmers at a price discounted by public funding. At least in Bahia, spraying the *Trichoderma* fungus appears to help to control witches'-broom and increase cacao yield, particularly when combined with the use of tolerant cacao varieties, adequate pruning of diseased plants, and copper fungicides. How well and how consistently this approach works will depend on the behavior of farmers, cacao prices, and, of course, the intricacies of the biology of cacao and the species on which it depends, many of which we do not yet understand. Here again, the ancient story repeats itself. A crop is moved from its native region and thereby escapes its enemies, only to be reunited with them again, which precipitates the search for enemies of the enemies. Meanwhile, our ignorance of the natural history of cacao continues to limit our ability to produce cacao fruits and, ultimately, chocolate. Buried, as a result, in every chocolate bar is a measure of just how little we know.

How hard would it be to destroy essentially all the cacao in the world? This is the question the story of Timóteo most clearly answers. It would not be hard. It would take a bag full of infected branches. This reality is troubling in terms of the future of cacao, whether from agricultural terrorism or the far more common accidental introductions.

How many other crops might be as easily destroyed as cacao, whether by a rogue individual, a terrorist organization, or another government? Crops are easy targets because they are grown over large areas, but a pest or pathogen need only be introduced at one or a few spots. It is not hard to "weaponize" a pest or pathogen. Sometimes, as the case of Bahia's witches'-broom makes clear, all it takes is putting a bunch of sticks in a bag. What's more, it is hard to detect an act of agricultural terrorism until it is relatively late, and then

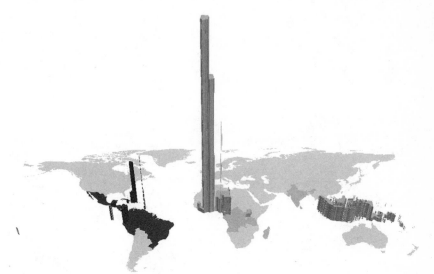

Figure 7. Global cacao production per unit area by country (2005–2014). Dark countries are those in which witches'-broom is now found. Light countries are those in which witches'-broom is absent—for now. Most of the chocolate you consume comes from cacao grown in West Africa. *Data source: FAOSTAT. Figure by Lauren Nichols, Rob Dunn Lab.*

once it is detected, it is hard to assign guilt. It is even hard to know for sure where a pathogen or pest might have come from. Many crop pathogens and pests remain to be studied and named. Of those that have been named, the vast majority have not been studied genetically in sufficient detail to identify where a new introduction may have come from. The United States does not even have a thorough or up-to-date list of the pathogens and pests that would have the biggest impact were they introduced. The list that does exist contains twenty species, whereas two thousand are possible problems.

7

The Meltdown of the Chocolate Ecosystem

In the aftermath of the tragedy of witches'-broom in Bahia, most of the cacao produced in the world is now grown in West Africa, on small farms ranging in size from a few trees to a few hectares.[1] The chocolate from these farms is the base of a $20-billion-per-year industry. It pours into chocolate bars and candies around the world. West African cacao is susceptible to witches'-broom and frosty pod rot. Neither of these pathogens has yet arrived in Africa. Eventually they will arrive, probably by accident rather than through an act of agricultural terrorism, but it will be an arrival all the same. Border controls can stall the inevitable for decades or even centuries. Yet even in the absence of these pathogens, West African cacao does not grow in the luxurious absence of pests and pathogens. It is besieged.

Early in its arrival in Africa, cacao was beset with problems not known in the Americas. One of these problems is a disease in which the leaves of the cacao trees redden and their shoots swell. This swollen-shoot disease eventually kills the trees and, in doing so, increases the price of every bar of chocolate you buy. It appeared in the 1900s, was controlled for a while, and is now back.

When swollen-shoot disease first appeared on cacao plants, in the early 1900s, West Africa was still divided among colonial powers—the British, French, and Portuguese. In the British colony of Gold Coast, where cacao trees were densely planted, hundreds of thousands of cacao trees died from the disease. Then as now, farmers in the region had small lots with a few cacao trees in the back to pay for things they could not grow (oil, matches, school uniforms) and cassava and other plants in the front for sustenance. The large companies that bought cacao to make chocolate could afford to ignore the problem; when cacao was lost in one region they would buy it from another (so long as there was another). For consumers, the chocolate produced from cacao is a luxury crop. For the farmers who grow cacao in Africa, its success is essential.

Colonial powers, vested in the success of agriculture in their colonies (but not others), sent scientists to deal with the increasing problem of plant pathogens on cacao, coffee, cassava, and other crops. Gold Coast was a British colony, and so a British scientist, Peter Posnette, was sent there from England.[2] He relatively quickly identified a virus as the cause of swollen-shoot disease—cacao swollen shoot virus (CSSV). Cacao swollen shoot virus is not from the Americas, where cacao was domesticated. It instead hopped from forest trees native to West Africa to cacao trees planted there. It is what plant pathologists call, somewhat euphemistically, a new encounter pathogen, a pathogen that newly encountered cacao once it was planted in great abundance in West Africa. It is one of several new encounter pathogens that attack cacao. The more of a crop that is planted in a region, the longer it is present, and the more poorly defended it is, the more new encounter pathogens will evolve the ability to colonize that crop.

All the cacao trees being grown in Gold Coast, most of which were of the Amelonado variety (just as in Bahia, Brazil), were susceptible to the virus. In order to deal with the virus and to improve West African cacao in general, Posnette cofounded what was then called the Central Cocoa Research Station, now the Cocoa Research Institute of Ghana, based in Tafo. Posnette and a team from the research station then traveled to Trinidad, where Posnette had spent two years studying the pollination of cacao. Once there, they crossed varieties from the upper Amazon of Peru with those from Trinidad through hand pollination.[3] They then brought the resulting seeds back to West Africa. The hope was to find or breed varieties that would resist swollen-shoot disease, whether initially or after even more breeding. It was painstaking work, demanding great patience. In the best of scenarios a new cacao variety, once it was formed as a seed, could be made to produce another generation in four years. Individual plants then had to be tested, one after another, for resistance to swollen-shoot disease as well as to other pathogens.

The resulting trees were not resistant; they were tolerant. They still suffered from the virus, just less than did other varieties of cacao. The new trees played a role in the economy of Gold Coast and played an even bigger role in helping to stabilize the economy after the region became Ghana, the first country to gain independence from colonial rule in West Africa. The reprieve, however, was partial and temporary. It was shortest-lived, some would later argue, in those places where tolerant trees were planted and pesticides were applied liberally, though why this was the case was not immediately clear.

Today West African cacao trees are once more dying from swollen-shoot virus. Virulent forms of the virus have spread throughout West Africa, and more than a hundred strains have been found in the region to date. Stopping the virus, or even just slowing it, will require understanding the complex ecosystem of which cacao trees are a part, a kind of ecological Rube Goldberg machine composed not of pulleys and tipping buckets of water but instead an incredible diversity of life, including fungi, flies, bees, and even wars among species of ants. If everything goes right, this miraculous ecological machine produces cacao beans, but right now everything is not going right. To start producing more beans again we need to understand each of the many elements of cause and effect in the system, especially those associated with the wars among ants.

. . .

Traditionally, cacao, like coffee, was farmed exclusively in the shade of the rain forest — rain forests filled with poorly studied species of ants.[4] The ants are key to the story of African cacao. In any given cacao plantation shaded by rain-forest trees one might find more than a hundred kinds of ants, many of them not yet named. On the ground are army ants, moving in great mass from place to place, eating everything small enough to be carried away. Beneath logs and the litter of leaves one finds trap-jaw ants, with mandibles that snap shut on prey faster than nearly any other animal can move any part of its body. Other ant species, specialists, eat only centipedes or snails. But the ants with the most consequence for cacao are those up in the trees, among their green canopies.

Canopy ants stake out territories that span whole trees or groups of trees and can reach abundances far greater than those of all of the other animals in the trees combined. Certain canopy ant species dominate certain trees or groups of trees, a pattern ant biologists refer to as the ant mosaic. But unlike tile mosaics, those of ants are always shifting as wars between ants tip various trees in favor of one or another dominant ant society. There was long a mystery in these mosaics. The ants were too abundant, more abundant than any predator could be, more abundant than their potential prey. It was as if the Serengeti were filled with hundreds of thousands of lions and just a few wildebeests. The solution to this mystery is that the ants living in the canopy are not predators at all but rather shepherds.[5]

Some of the ants in the canopies of rain-forest trees live on and care for mealybugs; they care for the mealybugs in much the same way that shepherds tend sheep. The mealybugs feed on tree sap, but they must drink vast quantities of it in order to get enough amino acids and other nutrients. As a result, they must also excrete huge amounts of sugar, their excess. This excess is manna for ants (the manna mentioned in the Bible was literally honeydew from insects), the sweet milk of their abundance. The ants not only consume the sugar, they also take care of the mealybugs in order to monopolize their supply. The ants build little tents over each clump of mealybugs that protect them from rain, parasites, and predators (and, as it turns out, pesticides). The ants then move the mealybugs around so that each is in just the right spot. In every rain-forest canopy in the world, the empire of ants is built on sugar, which allows ants to become far more abundant than if they lived by predation alone. The shifting mosaic of the canopy is not, then, a war for space so much as it is a war over the mealybugs and who might tend them. But there are exceptions, more details in this already complex story. Weaver ants (*Oecophylla longinoda*) achieve dominance through predation and by relying on scale insects, a distant relative of mealybugs, rather than on mealybugs themselves, for their sugar supply. This obscure detail turns out to be of great significance to cacao.

In West Africa, once cacao was introduced, ants moved in. Many brought their mealybugs with them, in great numbers. More than twenty different ant-tended mealybug species have been found on cacao in West Africa alone. This much was already known when Harold Charles Evans, a plant pathologist, arrived along with several entomologists in what was Gold Coast but had become Ghana, a newly independent country.[6] Evans arrived in Ghana with CABI (the Centre for Agriculture and Biosciences International) in 1969. His work until that point focused on fungi capable of living in extreme places. He studied, for example, the fungi living on coal waste.[7] He was and is very good at solving biological mysteries. He has an eye for the complex hidden in the simple.

Evans and his entomologist collaborators figured out that the shelters canopy ants make for their mealybugs are made out of bits of chewed-up cacao pods and other found materials. If this were the end of the story, it would be a lovely example of a mutualism. The shepherding ants protecting an insect on which they depend for food in such a way that both ant and mealybug benefit—using recycled materials, no less. But Evans noticed something more. Some of the ants were planting, in their care for the mealybugs, a sort of poisonous seed.

The cacao pods the ants use to make the shelters for the mealybugs often, Evans revealed, contain species of *Phytophthora*, relatives of the late blight that affects potatoes. These *Phytophthora* species infect the cacao pods.[8] The canopy ants, such as species of *Crematogaster,* that make shelters out of cacao pods appear to, in doing so, transport the *Phytophthora* around trees.[9] The ants spread the pathogen from branch to branch much in the same way that humans, in importing lumber, might unwittingly bring a pest from one shore to another.[10] In isolation, the shelters the ants build, laced with *Phytophthora,* are bad but not tragic. But some of the mealy-bugs cared for by the canopy ants also transmit the swollen-shoot virus (CSSV). When a tree has both the ants that make houses out of cacao pods and the ants that care for the mealybugs that vector the swollen-shoot virus, the virus and *Phytophthora* oomycetes together are worse for the tree than either is on its own. The virus makes it easier for the oomycetes to invade deep into the tree. When this happens, even trees that are relatively tolerant of the swollen-shoot virus suffer. This complex situation is the worst-case scenario for cacao (short of the arrival of witches'-broom)—what some call an ecological meltdown. It also appears to happen relatively frequently, and as if that weren't bad enough, if CCSV were to make it to the Americas it would likely find a willing mealybug there to vector it. *Phytophthora* is present in some cacao plantations in the Americas, but not the extremely virulent variety present in West Africa (a species known as *Phytophthora megakarya*). What, then, is our best hope for dealing with the swollen-shoot virus and *Phytophthora*?

. . .

Early on, Harry Evans and others suggested that one possibility was to enlist the ants' own wars in helping plants. While most of the dominant ants found in the canopies of cacao trees might be thought of as enemies of the cacao, the weaver ant (*Oecophylla longinoda*) and another species, *Tetramorium aculeatum,* are not. They are friends with great benefits. They do not care for the problematic mealybug species and so spread neither the swollen-shoot virus nor *Phytophthora*; they also eat the pests affecting cacao trees and actively fight other ants. Using them to control other ants has a long history. In Vietnam and China farmers traditionally carried the nests of weaver ants (made from leaves woven together with the silk produced by larvae) to orchard trees. Weaver ants are still used this way in some regions of these two countries, typically with great success. Yet as much as this approach long engaged the interest of scientists—and has precedent—it never caught on among those farming cacao in West Africa. It was too hard, perhaps, or too unpredictable (though it remains possible).

Another approach was to try to figure out a pathogen that might attack the problematic ants but not the beneficial ants. Evans worked on such an approach for a while. He began to study zombie fungi, which take over the bodies and immune systems of ants. He thought he might be able to use some of these pathogens to control the ants that shepherd mealybugs.[11] My enemy's enemy is my friend. Or in this case, my friend's enemy's enemy is my friend. As Evans studied these fungi he found that in the tall forests not far from the cacao fields the fungi that attacked both the ants tending mealybugs and the mealybugs themselves were diverse and abundant, and the ants and mealybugs were kept in check by the fungi. It was a specific case of the general principle that understory trees, when grown in high densities and sunny conditions, suffer from pathogens more than other trees do. A follow-up study exploring ways to favor such pathogens in cacao plantations, in addition to the use of shade trees, would have been useful. But any attempts proved too complex to be practical for farmers. Nonetheless, Evans's interest in these fungi would still play a role in attempts to improve the health of West African cacao—albeit a more indirect role.

In 2006, Evans was contacted by a young ant ecologist, David Hughes. Hughes is a pugnacious Irish ecologist, a lover of deep truths about the world regardless of whether they have any practical applications. Hughes wanted to talk to Evans about the fungi that attack ants. Such fungi and their evolution were Hughes's research focus at the time.[12] He wanted to know everything Evans knew about these fungi. He called Evans; he wrote him. In Evans's words, he badgered him. The two finally made contact and shared stories of ants and fungi. On some level, Evans must have seen in Hughes a kindred spirit, even though Evans's real focus was saving crops—he'd dedicated the decades of his professional life since working in Ghana to the control of plant pathogens—and Hughes's was basic biology.

Soon Hughes found himself on a plane headed to Brazil to collect fungi with Evans. Over time, Hughes and Evans began to travel together—to Brazil, Australia, and then Ghana. On these trips, Evans was always looking out for new pathogens that affect crops and new insights into old pathogens that might help farmers.

Hughes, meanwhile, was still searching for zombie fungi. He just wanted to understand how they worked and what they did to ants. Hughes, as he has said, is fascinated by the ways in which societies fall apart, be they societies of insects or humans. But as Hughes and Evans traveled together, Hughes's interests began to change: he began to worry less about what held societies of ants together and more about what threatened to tear those of humans asunder. Hughes and Evans started to consider together what one might do about the ants, fungi, viruses, and oomycetes in West Africa. Each of the ideas they have had together about how to deal with the pathogens of cacao relies on an understanding not just of one pathogen or another but also of the whole web of interactions in which cacao trees are enmeshed. In some ways, this complexity makes controlling the threats to cacao more difficult. But it also makes some simple solutions possible. Then a rather ordinary possibility presented itself.

A student working with Hughes, Megan Wilkerson, began to try to control cacao pathogens by means of an incredibly cheap process. She breaks open the little tents some ants make for mealybugs using soapy water. The soapy water disrupts the tents, kills some of the mealybugs, makes the survivors more susceptible to pesticides, and tends to disfavor the ants that tend mealybugs and favor those, such as weaver ants, that don't. Amazingly, it appears to be working. Soap. Water. Spray bottle. It's a simple technology that makes sense only in light of deep biological and historical understanding, and it has the potential to help every cacao tree. But even if it just saves a few cacao trees on a few small farms, it will make a big difference to those farmers—the difference between buying a school uniform or book and not being able to.

Similar approaches are now being tried with coffee. Coffee, like cacao, faces many threats. Among them are the coffee berry borer, a beetle that burrows into the seeds of coffee—where it uses a fungus to detoxify the coffee's caffeine and then eats to its tiny, simple heart's content—many other insects, and a rust, the same rust that destroyed coffee in what is now Sri Lanka in the 1800s, where Arabica coffee (*Coffea arabica*) was being grown. That rust appears to be nearly unstoppable. One early response was to shift to varieties of coffee that were resistant to it. This meant planting Liberian coffee (*Coffea liberica*), which was slightly resistant to the rust, but only temporarily. Then farmers switched to Robusta coffee (*Coffea canephora*), which was fully resistant but also relatively flavorless and more bitter when compared to Arabica varieties. Cheap coffee tastes worse than expensive coffee because cheap coffee is primarily Robusta. Yet, Robusta saved the West from having to join the British in drinking tea. Or, at least it did temporarily. The rust has continued to evolve, and now it is attacking essentially all varieties of coffee, including the Robusta.

One of the places where coffee is still doing well is Latin America. Early on, this was because Latin American coffee had escaped the rust as a result of its geography (that story again), but now rust is in Latin America (how it arrived, no one is sure, though it is likely to have come from Africa) and the coffee is still doing better. Why? The answer, at least in part, is publicly funded research. Some Latin American countries have invested in research institutes for coffee. When at their best, such institutes, whether in Latin America or elsewhere, have worked to combine the use of fungicides with the breeding of new varieties of coffee, be they from existing coffee varieties or by breeding with the threatened wild relatives of coffee. In addition, they manage the whole plantation ecosystem—its full diversity of useful and harmful species. Recently, this management has begun to include what may well, in the future, be a secret weapon, biological control.

In Latin America, especially where coffee is grown under shade, ants tend mealybugs on the coffee. Here, though, the story diverges from that of cacao. On coffee, the mealybugs are infected by a fungal pathogen. That unusual pathogen also parasitizes coffee rust. Where the mealybugs are more abundant, so is the fungus, and as a result the abundance of the coffee rust declines. In short, at least in terms of our current understanding, shade coffee favors ants that favor specific mealybugs that favor a fungus that, in turn, kills coffee rust. No one ever said this stuff was easy.[13]

. . .

Learning about the natural history and ecology of a crop in order to tip the natural interactions the crop has with other species in its (and our) favor is often called agroecology. It is a growing field whose existence suggests the ways in which we have ignored the natural history and ecology of our crops. If we tended to pay attention to the species with which our crops interact, exploiting those species and their abilities would just be part of crop science. Instead it is such a radical notion that it gets its own name. Agroecology is a hopeful way forward for many crops, especially in developing countries, where many farmers are poor. For cacao, as for coffee, agroecology simply makes more sense than the alternatives, including resource-intensive methods such as the large-scale use of pesticides and fungicides.

One could, for example, spray pesticides on cacao trees. This is what the intensification of agriculture often yields—a shift to general solutions that we hope will apply regardless of crop type, condition, region, or anything else. But the complex interactions of cacao with its biological neighbors make clear just how insufficient this kind of approach is. Farmers who grow cacao, like those who grow cassava, are unable to afford to buy pesticides and fungicides on a regular basis. More important, killing mealybugs with pesticides is very difficult; they hang close to the plant and cover themselves in a wax that is resistant to pesticides. Meanwhile, as those ants that tend mealybugs nest high in trees or in crevices and hence escape pesticides, the beneficial weaver ants nest low and in the open and are killed. A massive pesticide-spraying initiative was carried out in Ghana beginning in the 1950s. That program killed off some of the trees' pests, but they returned quickly and even in some cases developed resistance to the pesticides being used (i.e., a new pest evolved that had not been a problem before). The pesticide did, however, kill off populations of beneficial weaver ants, which, because of the long life and relatively slow growth of their colonies, were unable to recover after the spraying. It is argued that this spray program is the reason that cacao plantations suffered in Ghana in the 1960s relative to other countries where pesticides were not used in such large quantities. So as we have seen, pesticides can favor mealybugs, swollen-shoot disease, and frosty pod rot.[14] They kill beneficial ants and the parasites and pathogens that attack detrimental ants. But there are other problems, too.

Pesticides also kill the pollinators of cacao. And the production of cacao, the amount produced on any given farm, is tied to the abundance of pollinators. The biggest limitation in cacao production, in addition to that posed by pathogens and pests, is whether the cacao flowers are pollinated. Most cacao flowers are not pollinated, so the production of fruit in a cacao plantation is far less than it would be if each flower were fertilized. Cacao flowers in plantations are pollinated by flies—midges. These midges are, however, insufficiently abundant in most plantations to pollinate the cacao well. Their abundance can be increased if more leaf litter is left on the ground, because the flies have sex in among the rotting leaves, where the females lay their eggs. Their abundance is likely decreased when pesticides are used, though the abundance of the flies is always insufficient—perhaps, some have argued, because in its native range cacao is pollinated by a bee that did not move when cacao moved, though no one has ever studied the pollination of cacao in the wild.[15]

As with the study of life in general, we are far more ignorant than we tend to assume. The story of cacao, like that of cassava, is emblematic of how little we still know about our crops in general, particularly in the tropics. The best way to grow healthy cacao is probably still to plant trees in small patches beneath the canopies of rain-forest trees, the way the Aztecs did and, in such circumstances, doing whatever we can, be it soapy water or something else, to favor the species that help defend cacao against its pests and pathogens.[16] While doing so, we must save each and every variety of cacao so that the best variety might always be on hand for whatever new problem emerges, and to hope that the pathogen or pest over the hill does not arrive. We must also invest far more in controlling the arrival of plants and animals across our borders than we currently do. The border control that slows you down as you travel from one region to another buys us time in the agricultural sense and, in doing so, saves lives and billions of dollars. We must also do something else. We must study each and every species on which crops such as cacao depend (or might depend)—their predators, pathogens, mutualists, and everything else. We can map climate, globally, at a fine resolution. We can map soils. We can even predict a great deal about the future of climate and soils. But we do not yet know enough to predict the full set of species that might live in a particular place and how their webs of interactions affect our crops. We need to be able to map, in detail, the layers of life. At our current pace, we are centuries from such a map.

What would it take? We need studies of tens of thousands of species out in our coffee and cacao plantations, out in mango orchards, and in other places where plants are grown. We need to study hundreds of kinds of crops, each associated with thousands or in some cases hundreds of thousands of species. In my book *Every Living Thing*, I wrote about attempts in the 1980s to survey all the species in a single park in Costa Rica. When the funding fell through for that effort, attempts shifted to surveying all the species in the Great Smoky Mountains National Park. When the Great Smoky Mountains National Park inventory proved too difficult to complete, scientists, including those in my laboratory, turned to surveying backyards. We still haven't fully surveyed, much less understood, all the species in a single backyard anywhere in the world, much less the life in even one crop ecosystem.

To really understand all the species living with our crops, we will need to embark on one of the most ambitious projects ever undertaken in biology, an all-taxa inventory of the biodiversity of farm fields and orchards. It would need to involve tens of thousands of scientists and scientists in training. But it is a project that would have far greater impact than, say, the space program or efforts to reveal the detailed networks of the human brain.

8

Prospecting for Seeds

If science ceases to be a rebellion against authority, then it
does not deserve the talents of our brightest children.
                    —Freeman Dyson, *The Scientist as Rebel*

Potatoes fueled the economic success of Europe, wheat and corn
that of North America, and sweet potatoes that of Asia. These
crops were nearly all farmed using one or a few varieties, varieties
that grew bountifully in the absence of their enemies. But by the late
1800s the pests and pathogens to which those varieties were susceptible started to arrive in force. There was the potato blight, of course,
but also the potato beetle, and then wheat rusts, corn smuts, and so
many more maladies that no one could really keep track of them. To
continue to farm crops borrowed from other regions, the developed
world was going to need new varieties of seeds, and ideally they
would be available in advance, not just when some crop failed. The
question was whether anyone could actually go back and gather the
seeds that the conquistadors had missed.

In the early 1900s, crops failed as frequently in Russia as they
did in any temperate region. Those failures led, again and again, to
famine. Nikolai Vavilov grew up during several Russian famines.
His family was relatively affluent, but not so affluent as to be
shielded from bad years and the failures of crops. As a result, they
ate simple food, whatever was most available in a particular season
or what had been stored, fermented, from the previous season. Vavilov's father, a textile merchant, encouraged Nikolai and his brother

to follow him into business. It was a path to success in a country that historically had few such paths. But times were changing, and the old paths to success were no longer the only ones. Vavilov's generation believed in the power of science to help transform society. His brother became a physicist, one of his sisters a microbiologist, the other a doctor. Nikolai, in turn, studied at the Moscow Agricultural Institute. In doing so he embarked upon a professional life focused on plants, agriculture, and food. Just what he would do in such a career, and whether it would be anything other than ordinary, was not obvious at the time.

Vavilov graduated from the institute in the spring of 1911, having focused on breeding varieties of oat, barley, and wheat that were resistant to pathogens.[1] As a student, he learned about the work of Charles Darwin, whose *Origin of Species,* published in 1859, revealed the process of natural selection by which nature, red in tooth and claw, winnows the fittest form from all others.[2] To young Vavilov, Darwin's insights seemed both revolutionary and useful. Vavilov would come to think of himself as a student of Darwin; he always kept a portrait of Darwin in his office. Vavilov had also learned about the even more recent work of Gregor Mendel, an Austrian monk. Mendel, who died in 1884, developed the basic theory of the laws of inheritance traits—be they the fuzziness of pea pods or their resistance to pathogens. Regarding the experiments of Mendel and those who built upon them, Vavilov wrote, "The recent experiments in genetics have unveiled much more opportunities than a researcher of the past could...dream about."[3] At the time, the informal history of plant breeding stretching back some ten thousand years or more focused on strategies enabling farmers to choose, in each generation, the individual plants that had the traits they desired or at least were the closest match available. Darwin's and Mendel's insights suggested another model, one in which a scientist could systematically cross particular varieties of crops so as to produce offspring with, say, the fast growth of one variety and the resistance of another.

Vavilov wanted to understand how crop varieties were related to each other, how they varied around the world, and where one might find crop varieties with the most extreme traits—the strongest resistance to a pest; the highest tolerance for cold or drought. Such information seemed key to modern plant breeding. Vavilov wrote to professor Robert E. Regel, at the Bureau of Applied Botany, in Petrograd (now Saint Petersburg), expressing this interest. Subsequently Regel mentored Vavilov during the fall of 1911 and spring of 1912 at the bureau. Vavilov then returned to Moscow, where he organized a series of seminars. In one of these seminars, Vavilov gave a talk in which he argued for the value of the theory of genetics to agriculture.⁴

By 1912 Vavilov had done practical work on plant breeding and resistance and had intensely studied plant pathogens, plant taxonomy, and plant geography. Then in 1913, he was offered the opportunity to tour western Europe to visit the labs of scientists working there and, in doing so, learn about the cutting-edge research in plant breeding and genetics going on outside Russia.

Vavilov left Petrograd carrying letters of introduction from Regel; the letters became tickets into the laboratories of some of the most esteemed scientists of western Europe. He visited the lab of William Bateson, by then director of the John Innes Horticultural Institution, in Merton, and Reginald Punnett and Rowland Biffen at Cambridge. It was Bateson who had coined the term *genetics* and helped rediscover the work of Gregor Mendel. Bateson advocated a new way of understanding and working with all of life. With Bateson, Vavilov saw modern genetics being applied.

As he traveled, Vavilov was starting to envision the future. He believed Bateson's argument that generations of scientists would use genetics to breed new crops in ever more controlled ways; it was an idea he had held even before arriving in England, one that seemed forcefully validated by Bateson. But he also knew that if this was to happen it would require that scientists have access to the fullest possible diversity of genes and traits engendered in crops over thousands of years of human and agricultural history—thousands of years of selection of just the right crops for each region and culture on earth. Vavilov foresaw that as the tools of genetics improved, the only limit to human ingenuity would be whether or not geneticists had access to the full diversity of existing varieties. In October of 1914, just after World War I erupted, Vavilov was forced to return home. He arrived in Russia having collected both new knowledge and, from any place he could, new seeds. He made it home safely, but his collection of the seeds of western Europe was on a ship that hit a mine. The ship exploded. The collection disappeared. This was to be Vavilov's first experience — but not the last — with the challenge of collecting and saving seeds.

Vavilov continued his work back at the Moscow Agricultural Institute. He studied Russian wheat varieties and their resistance to various pathogens. He also began to work with younger researchers, training them and, as he did, expanding the scope of his projects. His studies included work on an unusual Persian (Iranian) wheat resistant to powdery mildew.[5] This Persian wheat could be bred with other varieties of wheat to produce new varieties that were both resistant to powdery mildew and high-yielding. But, he thought, it would be useful to get more samples. In 1915, he took brief trips to the Trans-Caspian region and Turkmenistan, which was adjacent to Persia and so might also be a place where Persian wheat was farmed. He found many new varieties of crop plants, but no Persian wheat.

Then after two years of steady work came a big opportunity. In May of 1916 the Russian ministry of defense contacted the Moscow Agricultural Academy. The ministry of defense had a problem of a botanical nature: "Could someone come help?" The academy recommended Vavilov for the job. The next morning Vavilov stood out in front of his house, eager and self-assured, wearing a gray woolen suit, a white fedora, and carrying an enormous backpack.[6] He had plant presses, too, and a working knowledge of the plants of the region — those that were known and those whose presence he could infer. He was an agronomist reporting for duty, eagerly.

Vavilov was told that many of the soldiers stationed in Persia at the time were getting sick when they ate bread made from the local wheat. The bread seemed to make the soldiers drunk and hallucinatory. The military wanted Vavilov to figure out what was happening. Vavilov had a good guess as to what the problem was. He probably could have recommended a solution in a letter,[7] but why would he? He was eager to get to Persia to collect more plants.

Vavilov was especially interested in finding new varieties of wheat and barley. He, like Biffen at Cambridge, was making good progress in breeding new wheats but was limited by the varieties he had to work with. Wheat sustained Russia and much of Europe, but just as conspicuously often failed during years when pathogens wiped out the crops or in regions such as much of the Eurasian steppe, where most of the time the short seasons and cold weather prevented Russian wheat varieties from growing at all.

When Vavilov arrived at the camp where the troops were getting sick, near Ashkhabad, he quickly diagnosed the cause of the illness. His guess had been right. The soldiers were accidentally consuming the seeds of a weed called darnel (*Lolium temulentum*), which were inadvertently being baked along with the wheat grains in bread. The darnel seeds were similar enough to wheat seeds to be gathered by accident.[8] The weeds themselves were not toxic but were home to a fusarium fungus, ergot, that lives as a partner of the plant within its stems and seeds. Ergot produces lysergic acid diethylamide, a.k.a. LSD. Hot bread contaminated with the weed and its fungus was making the soldiers high when it was eaten immediately after baking. The soldiers were told to not eat bread made from local grain, and the problem ceased. Pleased with his ability to solve the great military mystery in a day, Vavilov traveled on to the east.

Vavilov took his three horses and began to explore Persia. En route, he found a new wild perennial flax ripe with seeds, which he gathered in abundance. A new rice. New wheat varieties (though not the disease-resistant wheat he thought he might encounter). And more. He sped down the path, grabbing, studying, questioning. He was collecting new varieties and starting to get a sense of the regions in which farmers had bred the greatest diversity of wheat.

While collecting, Vavilov stumbled upon Russian Cossacks marching toward the Tigris River. Biologists in the field often look suspicious to authorities. Their behavior seems unusual and irrational, their equipment vaguely military, and their explanations ("I'm collecting plants to save humanity") beyond belief. In this Vavilov was no exception. He was taken immediately to the guard post. He was questioned, and each of his answers made the Cossacks surer he must be a spy. Vavilov's first language was, of course, Russian, but he was taking notes in English, as he had done since his trip to the UK. This was odd and suspicious. More suspiciously, he had with him books written in German.

Vavilov and his team were passed along from the guard post to another authority, this one charged with the specific task of exterminating German "vermin." The vermin detector was given one thousand gold rubles for every German spy he found. With that incentive, and considering Vavilov's strange behavior, the detector felt he had clearly hit the jackpot. Fortunately, Vavilov was freed after three days, having talked his way to freedom.[9]

Upon his release, Vavilov kept going east, another thousand kilometers on horseback. He had planned to return home to Moscow in August. But he decided on another plan. He would instead trek up along the border between Turkmenistan (then part of the Russian Empire) and Afghanistan toward the high plateaus of the Pamir Mountains, at the intersection of Tajikistan, Afghanistan, and China. It was in these plateaus that Vavilov would make the biggest discoveries—though not until after he was chased by a mob during an uprising. He narrowly escaped the mob only to be arrested, and then, once more, he was, very fortunately, released.

Beyond the political and social challenges Vavilov faced, his travel to the Pamir Mountains was physically arduous. It would have been hard at any time (especially after having already logged several thousand kilometers with a growing collection of dried seeds and plants in tow), but because he was traveling very near to the line that separated Russia's troops from those of Turkey—one of the front lines in World War I—he could not take the normal route. He would have to go, in the middle of winter, up and over glaciers to get to the plateaus. He was by then traveling with six horses, two porters, and a local guide, Khan Kil'dy Mirza-Bashi, who taught Vavilov about local crop varieties, translated for him, coordinated logistics, and, more than once, saved his life. They walked along a path, six feet wide at its widest, at the edge of a cliff above a valley that dropped hundreds and then thousands of feet below them. Vavilov found totally unknown varieties of wheat, rye, peas, and lentils, many of them resistant to powdery mildew. Then, he found a wheat that ripened earlier than any he had yet seen. It was perfect for use in the cold, dry, northern reaches of Russia, where seasons were short and wheat varieties needed to be able to take advantage of as many days of sunlight as possible. With this wheat, a variety highly valued in the Pamirs at the time (as it is today), Vavilov might expand agriculture north in Russia and save lives—or he would, anyway, if he ever went home.[10]

On his grand journey from Persia to the Pamir highlands (and, later, on each of his many subsequent trips to Africa, the Americas, and Asia), Vavilov saw particular plant varieties of interest, but he also saw broad patterns, the rules of both nature and human societies. For example, while an individual village might farm relatively few kinds of peas, wheat, and corn, in traveling from one village or region to the next, one could find thousands. This seemed to be especially true in mountainous regions, where plants might vary not only with climate (up and down the mountain) but also with longitude and latitude (in one valley relative to another similar valley). Some aspects of the diversity of crops seemed to him to be predictable: he was revealing, it seemed, the mysteries of the evolution of the diversity of crops. How many more varieties of crops might exist, hidden in one or another village? No one knew. Even the varieties known to scientists had never been gathered systematically in one place.

. . .

Vavilov returned from his odyssey in the spring of 1917, the year of the October Revolution. He was just thirty years old and had not yet defended his PhD (that would come in 1918). But he had already made a range of important discoveries, on the basis of which he was appointed professor in the department of agriculture and plant breeding at Saratov University. There he set up an experimental research station that became the Saratov branch of the Bureau of Applied Botany. At the time, the Saratov branch was essentially the only place in Russia where plant breeding was actively taking place. By virtue of their accomplishments, and despite the vicissitudes of war and revolution, Vavilov and his students had become responsible for the future of food in the biggest country on earth. This was partly because of the paucity of work being done elsewhere in the country and partly because of the ambitiousness of Vavilov's efforts. In 1918, Vavilov's mentor, Robert E. Regel, felt comfortable writing that although many had studied plants' resistance to pathogens and pests (immunity, he called it), no one had ever approached the problem "with such a breadth of views and comprehensive coverage of this problem, as Vavilov has done."[11]

The work of seed collection and breeding new crop varieties Vavilov was leading at Saratov was a work of contrasts. On the one hand, it involved Vavilov's adventures, his travels, his diligent attempts to learn everything he could from each farmer he met, no matter where that farmer lived and no matter what language he spoke. On the other hand, it also required the more monastic duties of saving the seeds, growing them out, documenting their details, and curating them carefully, all of which took time, patience, and a mind-numbing diligence.

The more Vavilov's endeavor expanded, the more monklike work there would be. The more, too, there would be the need for exploration. And his work expanded a lot. In 1920, after Regel died from typhus, Vavilov became the head of the Bureau of Applied Botany[12] and moved to Petrograd with twenty researchers. By 1920, he employed sixty researchers. By 1921, things were going so well that Vavilov set up a division of the bureau's department of applied botany and breeding at 136 Liberty Street in New York City, a division that would find and buy seeds from the Americas. This division alone would add more seed varieties to the Russian seed collection than existed in the years before Vavilov started his work. Vavilov would use these varieties for breeding and to understand the evolution of the diversity of crops on earth. He was, some had already begun to say, "the Mendeleyev of biology."[13] This comparison was not, however, perfectly apt. Mendeleyev brought order to the elements. Vavilov wanted to bring not just order to the diversity of plants but also, based on that order, an understanding of ways to improve agriculture in the future in light of its history.

Vavilov could have done great work with only the seeds he brought back from Persia and those being supplied from North

America, but he wanted far more. He needed to keep collecting. But where? As of 1920, much of Vavilov's work had focused on wheat, barley, and rye, and in considering these plants he documented their diversity, cleared up ambiguities in their nomenclature, and tried to understand their evolutionary history. He not only unraveled aspects of the particular stories of these crops but also discerned general features—rules, really—of the domestication of crops. He used his understanding of these rules, when coupled with detailed study of the literature on various domesticated plants,[14] to begin to predict which regions would be most productive for study of each kind of domesticate. He identified geographic centers where farmers working with wooden plows, clay tools, and their hands seemed to have engendered or moved a disproportionate number of crops and varieties over millennia, crops perfect for the conditions—the climates and cultures—in which they lived. He called these regions centers of origin, with the dual implication that they were not only the locations where the greatest diversity existed but also the locations where that diversity arose. For example, the diversity of barley and wheat was greatest in Asia Minor, and hence it was in that region where this diversity likely arose. The diversity of potato varieties was highest in the Andes and Chile, and so this, too, was a center of origin. The diversities of chilies, cacao, tomatoes, and corn were highest in Mesoamerica, and so that region was their center of origin. And so on. Breeding crop varieties took time, and farmers had the most time to breed new varieties near where a crop was first farmed. He published a major work on this idea, "The Centers of Origin of Cultivated Plants,"[15] in 1926, but long before then he had begun to use the germ of this theory to guide his exploration.

. . .

On the basis of his centers of origin theory, Vavilov could predict where one might find new kinds of crops in the greatest numbers. In doing so he could finish the work the conquistadors had started, gathering the seeds most likely to be of use rather than simply those that traveled most readily or seemed most appealing to a hungry traveler. Repeatedly he left what was by that time the Soviet Union

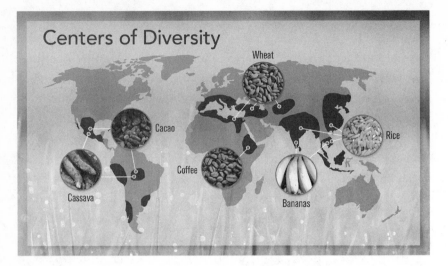

Figure 8. Crop centers of diversity, with exemplar crops and their centers of diversity in circles. *Figure by Neil McCoy, Rob Dunn Lab.*

for those remote regions. On each trip he took risks. On each trip he came back with riches—camel loads of seeds, elephant loads of seeds, horse loads of seeds, boatloads of seeds. He went on expeditions, on every continent except Antarctica, in sixty-four countries.[16] On each trip, he found more seeds; after each trip, his institute, which in 1930 had come to be called the All-Union Institute of Plant Industry (now the N. I. Vavilov Institute of Plant Genetic Resources, or VIR), would grow larger and more ambitious.[17]

We think of modern humans as destroyers of diversity. We humans do battle against the dark worlds of pathogens and pests. We killed the mastodons and mammoths and then, when they were gone, hacked at the wolves and bears. But farmers in villages around the world also created; they chose wild plants and carefully, by replanting favored forms in just the right ways, created more diversity—more diversity than nature offered on its own. They created thousands, perhaps hundreds of thousands, of new, useful crop varieties. Most crops in the world depended on an individual culture, village, or even just a family who tended to them. Every species of crop you eat, every variety within a species, has a history linked to villages, people, and places that is mostly lost to the anonymous grind of time. As the world became ever more connected and industrialized, and as local knowledge and farming approaches gave way to global approaches, these varieties would, Vavilov knew, be lost. This is partly why he often chose to travel to difficult-to-reach destinations. In places that were easy to get to, many of the most interesting folk varieties were already gone.

In the early 1900s, as populations around Russia and around the world grew, the need to produce more food on a finite amount of land was increasing. Politicians noticed. The hungry noticed. Vavilov himself had lived through multiple famines. In central Russia alone droughts would occur in both 1920, when Vavilov was still in Saratov, and 1924, by which time Vavilov had moved to Petrograd. There would be another drought in 1936. It was becoming easier to envision a time when the whole world might be similarly susceptible.

Vavilov's work was lauded and rewarded by Lenin, who spoke of a society that offered opportunity for artists, scholars, and scientists such as Vavilov. His work was also lauded by society at large and by other scientists. Soon Vavilov employed researchers and staff all across Russia. Some were stationed at the core seed collection in Leningrad (the former Petrograd). Many others, though, were at the thirty-six plant breeding centers across the Soviet Union. At these sites, Vavilov and his team carried out the breeding of many varieties of crops. They then tested the most interesting of the new varieties against the diverse climates of Russia through use of 115 sites. The stations were located so as to cover each of the climates of the Soviet Union, to maximize the number of varieties from Vavilov's collections that could be grown.

In 1925, at the age of thirty-eight, Vavilov was elected a corresponding member of the Academy of Sciences of the USSR; five years later he became the youngest person to become a full member. He was, he believed, on track to build a cathedral of science out of which future innovations in seeds and crops and sustenance would come. In his near-religious enthusiasm for this project, Vavilov had gathered a community of men and women around him who shared not only his belief in the power of plants but also his determination to build something dedicated to civilization that was far more ambitious than any of them might achieve in their lifetimes. He had, he told one reporter, "a hundred year plan."[18] Vavilov even had a plan for the order of things. He started with wheat, barley, oats, and cabbages. He would move next to melons and strawberries.

Vavilov knew that in order to sustain the species we depend on, we—the collective we, the big human we—would ultimately need several things. We would need to save the knowledge and varieties produced by traditional farmers around the world. We would need to understand the history of those varieties and the people associated with them. We would need scientists and farmers who tinker with those varieties in light of particular problems. This tinkering, of course, includes the production of new hybrids, a process that is nothing more than what traditional farmers have long done but informed by the insight that the science of genetics gives us. We would also need new sorts of tinkering—genetic engineering, some call it—which speeds up the process. Finally we would need an understanding of the basic biology of the forests and grasslands in which traditional crop varieties and their ancestors evolved. Vavilov made each of these points and, one can argue, hoped that these research agendas would be part of an even bigger plan, a five-hundred-year plan—after he completed the first hundred years of work, gathering the seeds.

By 1935, Vavilov and his team had gathered between 148,000 and 175,000 varieties of crops and their wild relatives.[19] Each variety contained the story of the people who'd bred it from its wild stock or from other domesticated seeds. Back on the farms, the stories of these varieties, of course, continued, unfolding in each place he had gone as new crops were traded, favored, and created. From this collection, new forms could be systematically bred, forms that would complement whatever ingenuity continued in farmers' fields. Vavilov, of course, did not gather all the traditional knowledge associated with each crop, though he tried to gather as much as he could. Nor did he gather the species on which each crop depended—the microbes associated with its roots, the bees with its flowers, the predators that controlled its pests. Nor did he understand all the connections among these species. The seeds, though, were a first step. Then came World War II. By the time the war began, Vavilov had disappeared, and his collections, his precious collections, were under siege.

9

The Siege

The greatest service which can be rendered to any country is
to add a useful plant to its culture, especially a bread grain.
—Thomas Jefferson, "Summary of Public Service"

Man shall not live by bread alone.
—Matthew 4:4

On June, 22, 1941, Hitler's soldiers crossed the Soviet border. By
September, they had arrived at the edge of Leningrad. Once
the city of the czars and their palaces, Leningrad was, in 1941, the
cultural heart of the Soviet Union. It was the center of music, art,
and science. Initially the goal was to take the city, but in a last-minute
change of strategy, the Germans decided on a blockade. They calculated that it would take just weeks—or, at maximum, months—to
starve out the city. After Leningrad, they would take on the rest of
the Soviet Union.

By late September the Nazis had severed all Leningrad's connections to the world except the remotest paths and, once winter settled
over the city, the wide route over Lake Ladoga up into the Ural
Mountains. No fuel or food could get in. Little could get out. Young
men by then had already gone to the war front, leaving Leningrad a
city of children, those too old to fight, and women, all of whom
began to stockpile any resources they could. At the time, a few million people lived in Leningrad, and among them are an equal number of stories about what they tried to save during that autumn. In

this cacophony of tragedy, one story stands out—the story of the people protecting Nikolai Vavilov's seed collection.

Hitler created a special SS commando unit, the Russland-Sammelkommando, led by Heinz Brücher, whose only goal was to find and take Vavilov's seeds both from Leningrad and from the research stations across the Soviet Union.[1] Hitler's head of the SS, Heinrich Himmler, sought to settle Germans in much of the western Soviet Union and Poland. Brücher believed that Vavilov's seed collections would be necessary in order to make this conquered land— so different from Germany—productive.[2] Those running Vavilov's institute had to save the collection not only from bombs but also from Brücher and his unit.

The seeds at most immediate risk were not seeds at all but potatoes. In the Andes, Vavilov had collected thousands of pounds of potatoes of more than six thousand varieties, potatoes with potentially immense value to the Soviet Union (and the world). But potatoes store poorly as seeds, and seed potatoes, like those used for planting in Ireland, store worse. So at the time, the best option was to plant the potatoes and harvest them anew, which Vavilov and his team did each year at Pavlovsk, an experimental station located at a former residence of the czars around thirty kilometers (eighteen miles) southeast of Leningrad. But early on in the blockade the Nazis had begun to bomb the outskirts of the city, including Pavlovsk. The field in which the potatoes were planted was under fire.

Abraham Y. Kameraz and Olga A. Voskresenskaya, the de facto leaders of the collection, raced to gather the potatoes into boxes in the fields, doing their best to include at least some of each variety. They ran beneath the crowns of orchard trees that in the spring would be fragrant with the blossoms of hundreds of varieties of cherries, plums, and apples. They ran between artillery shells to pick them one by one out of the cold ground and put them in the boxes. They then asked the Red Army soldiers, who were themselves convinced of the value of the potatoes, to help transport the boxes. The soldiers helped pick up the boxes and drove them in military trucks to Saint Isaac's Square, in Leningrad, where the most complete collection of the seeds had already been gathered. This work of moving potatoes and other samples continued, furiously, until the Nazis took Pavlovsk.

On Saint Isaac's Square the seeds and potatoes were stored in the dark recesses of a building at 44 Herzen Street. The workers who remained with the collection decided to create duplicates of the most valuable crop varieties, which would then be moved. But which were the most valuable? The workers decided quickly, as best they could. But even as they made the duplicates, they knew they faced a second problem: how to evacuate so many seeds now that the city was completely surrounded. The whole collection—more than one hundred thousand samples and five tons of seeds—could be moved, it was thought, one small parcel at a time. A plan was hatched to transport some of the collection hundreds of miles across Lake Ladoga, once it froze, to the Ural Mountains. Other seeds could be given to people to carry in their pockets or hand luggage as they evacuated the city along the last open routes, though this obviously offered no salvation for potatoes. Duplicates of a larger number of seeds were packed into boxes with double walls that would be transported by train. The samples made it to a train car, but it was too late. After six months on the track, waiting for a time when it was safe to move—which never came—the train car was unpacked and the seeds brought back to the collection. Some seeds made it out, but few. The Nazis had surrounded the city too completely, and winter was coming fast and hard. The scholars in the seed bank would just need to save the seeds where they had been consolidated, at 44 Herzen Street. The men and women there would spend as long as they had to on their singular mission.

At first the scholars were guarding the collection from the Germans—and with them the Finns, who were aiding in the blockade. But as the fall of 1942 turned to winter and food became scarcer, it became clear that they were also preserving the seeds from their fellow Russians, Russians who needed food. The government of Leningrad calculated that just a thirty-day supply of food remained and so allotted its manual laborers 250 grams of a mixture of bread and bran per person per day; just 125 grams per day was allocated for everyone else, including those working in the collections. That was to be the entire daily diet of millions of people. Hunger stalked the city.

Hunger meant that the collection was threatened not only by the physical conditions—the deadly cold and the equally dangerous wet—but also by the threat that hungry citizens of Leningrad might, in desperation, eat the seeds. While the collection represented a repository of genetic history, it was also, more simply, food. In the building at 44 Herzen Street were tons of rice seeds and wheat—which, of course, are simply rice and wheat grains—potatoes, and so much more. Break-ins started to occur once desperate people realized what was inside. The collection smelled of fruits, berries, and grains. A human could smell the richness of the collections through the walls. So could the rats.

The rats and mice of Leningrad are said to have been even more numerous in the first winter of the siege because humans started to eat the city's cats. Like the humans, the rats and mice were starving and cold, and they found the collection. They began to chew through the paper and wooden containers housing the seeds. The only thing that could be done was to further increase the protection around the seeds. Seeds in wood or paper were moved to metal. Metal containers were wrapped as tightly together as was possible. In eighteen rooms, the seeds were cared for by hand, one seed and one box at a time. The windows had been blown out by the bombings and boarded up, so the rooms were dark. Without electricity this meant that all work had to be done by the flicker of kerosene lamps. Finally, when the workers were convinced that the rats and mice were gone from each room, the rooms were sealed. The seals on each room were checked each day. Three to five workers staffed the building twenty-four hours a day. Eventually they even barricaded themselves in, neither coming nor going.

Outside, as of February, several hundred thousand Russians had already starved to death. The ration was no longer bread but instead malt flour, cellulose, and calfskins, foodstuffs on which no one could survive. Over its nine-hundred-day duration, the siege would ultimately claim 1.5 million Russian lives. Those who survived the hunger often died of cold (winter temperatures reached minus forty degrees Celsius, which also happens to be minus forty degrees Fahrenheit). No heat existed in the city; both coal and wood fuel were gone.

The scholars, too, were cold and hungry. What made their plight unusual was that while those outside had little access to food, the workers in the collection were surrounded by it. Their bodies were malnourished in a room filled with sustenance. At one point it became clear that, to be saved, the potatoes that had been rescued and were being stored in the cold needed to be reburied in the soil. The potatoes had already received so much extra care. They were rescued from Pavlovsk. They were also kept warm with a small fire in the stove even while the workers couldn't afford to use any wood to heat themselves and their families. If the potatoes froze solid, they would die—which, of course, was also true for the families of the workers. The workers might simply have consumed them instead of replanting; history would have forgiven them. They did not. Yet as their hunger grew, the scientists began to require of each other that no one go into a collection room on his or her own. They must always have company, so that no one might be tempted.

Meanwhile, Heinz Brücher coordinated the seizure of Vavilov's seed collections from some of the experiment stations. That this had not already happened in Leningrad was a sort of good news. Also good news was that the collection had not been bombed. The workers in the collection breathed a sigh of relief each time a bomb failed to hit them. Contrary to hopes that the Germans would retreat, and with them the siege, the siege continued throughout the summer. It continued for another year. It would continue until the spring of 1944.

By the winter of 1942, no one working in the collection was healthy any longer. Dmitri Ivanov was in the poorest condition. Ivanov was one of the oldest. He had been directly trained by Vavilov and had, with Vavilov's help, assembled the largest collection of rice varieties ever to exist, rice that could grow anywhere. Ivanov was starving to death amid this rice. But he would not eat it, even as the winter progressed. With time, he grew so hungry and gaunt that his body could take no more. In early January he died sitting at a typewriter, surrounded by bags of rice grains.

Ivanov's death was followed by those of others. Alexander Stchukin, a peanut specialist, died at his desk. A man named Gleiber, the keeper of Vavilov's field notes, died of starvation surrounded by those notes. Soon afterward, Georgy Krier, the man in charge of herbs, also died. Then Liliya Rodina, the oat specialist, was gone, followed by A. Malygina, A. Korzun, N. Leontjevsky, M. Shcheglov, and G. Kovalevsky as their names are recorded in Russian histories. All told, more than thirty of the workers from the VIR who remained during the siege would die. Presented with a choice between saving their own lives and saving Vavilov's great collection, they all chose the collection. They died, they believed, that some future generation might live. They died on behalf of a greater project, a great cathedral of knowledge and plants whose legacy would, they hoped, far outlast them.

· · ·

As for Vavilov himself, he was missing. His disappearance had occurred two years earlier, in 1940. He'd been out collecting plants in the Carpathian Mountains in the Ukrainian region of the Soviet Union when on August 6 a black car with tinted windows pulled up. Vavilov was asked to get into the car: "Stalin's orders!" He obliged. The car drove away, down and out of the mountains. That was the last Vavilov's friends had seen of him as of the winter of 1942, when they sat guarding the collections of seeds.

Vavilov had been arrested for collecting plants before, and each time, reason, justice, or both had prevailed. But Vavilov no longer lived in a right and reasonable country. He lived in Joseph Stalin's Soviet Union.[3] Vavilov was the most vocal and well-known scientist in the country, and his work proceeded from a modern understanding of both genetics and evolution. This put him in opposition to the scientist Stalin had chosen to be his right-hand man when it came to all decisions about science, Trofim Lysenko. By extension, this put him in opposition to Stalin, and no one stood in opposition to Stalin for very long.[4]

Soon after his capture Vavilov was imprisoned in the People's Commissariat for Internal Affairs, accused of espionage and of supporting the "false" science of Gregor Mendel and Charles Darwin. He was, in effect, accused of acknowledging what biologists call the modern synthesis (the coming together of Darwin's insights about natural selection and those of Mendel about the rules of inheritance). Vavilov was a revolutionary, and truth was the revolution—truth as elaborated through the study of rare crops, genetics, and evolution. Vavilov was the nation's most vocal advocate for science, a man celebrated around the world for his achievement. Surely if he had the chance to argue, Vavilov could convince those in power that he was doing nothing radical, that his work was necessary for the Soviet empire. But Vavilov was up against enemies that did not necessarily obey the laws of reason: the Soviet Union, Stalin, and the agriculturalist Trofim Lysenko.

Lysenko was a farmer who had shown that one could improve the yield of some wheat varieties by shocking them with cold, convincing the seeds that they had been through winter. This approach, called vernalization, did help increase the number of seeds able to begin growing early in the spring. It was a way of mimicking winter in a controlled way. But as it is said in Russia, when a man has a new hammer, everything looks like a nail. In 1933 Lysenko became codirector of the All-Union Selection and Genetics Institute, in Odessa, where he proposed to train a generation of "barefoot scientists" who would improve crops without worrying about what Lysenko considered to be the nuisance theories of genetics or natural selection. They could make seeds cold, and they would grow better in the cold; make seeds hot, and they would grow better in the heat. Lysenko came to argue that this approach was all that was necessary to improve Soviet agriculture. Plant breeding, genetics, and evolution were not necessary, nor was collecting seeds from far-flung environments. With his homespun story of becoming a scientist through work in the field rather than through formal education, and with his belief in improving crops through hard work and inspiration alone, Lysenko became the emblem of Soviet agriculture.

Lysenko held up his vernalized seeds and called for the end of genetics, much as a US senator might hold up a snowball and announce climate change to be a sham. In rewarding Lysenko with the position at the Odessa institute, Stalin had indicated clearly the direction in which he wished Soviet agriculture to go. Over time, Stalin's government offered an increasing amount of support for Lysenko.

At first Vavilov, who appreciated the initial contribution made by Lysenko's vernalization work, was supportive, too. Gradually he began to realize the extent of the threat posed by Lysenko, but he still seemed unable to believe it was real, much as scientists today struggle to come to the terms with the threat posed by the antiscience rhetoric of our leaders. He confronted Stalin, trying to present him with his science, trying to show him the scientific method, the beauty of genetics, and the flaws of Lysenko's approach. Stalin was furious. He screamed at Vavilov that his excursions overseas were a waste of resources, that he should instead be spending his time learning from the "shock workers in the fields," who were to shock the seeds with cold so as to improve them for the future. If Vavilov did not follow these instructions willingly, he would be made to follow them in other ways, as would all of Soviet agriculture. Lysenko, to varying degrees, would remain the unaccomplished hero of Soviet agriculture for decades. No one dared oppose him—no one but Vavilov.

On the first night Vavilov was imprisoned, Alexander Khvat came into his cell. Khvat was the investigator charged with extracting an admission from Vavilov, an admission that through his science he had become an enemy of the government, an agricultural spy on behalf of the British. Khvat started promptly upon entering Vavilov's cell at 11:35 p.m. and did not stop until 2:30 the next morning. When Vavilov did not admit to the crimes of which he was accused, Khvat came back the next day. And the next. Khvat or one of several other interrogators visited Vavilov four hundred times— nearly two thousand hours of interrogation and torture. Vavilov had done nothing wrong, and so he had nothing to admit to, nor would he admit to things he had not done, at least initially. But Vavilov was no longer whole. He was sick. He was no longer able to stand. His fellow prisoners had to help him remove his boots from his swollen feet and roll him onto his back, where he could collapse.

Eventually he broke. On August 24, 1940, Vavilov falsely confessed to being "a member of a right-wing anti-Soviet organization," the Peasants Party of Labor."[5] He invented the organization in the hope that the fabricated admission would stop his torture. It did not. Vavilov was tortured more. He was tortured until he confessed that his fellow scientists, his friends, were also members of the same party. The one thing Vavilov did not renounce, even after he gave up his friends, including Leonid Govorov and Georgy Karpechenko, was his belief in science. In many hours of interrogations, Vavilov continued to deny the torturers' assertions that modern science, Mendel and Darwin, were wrong.

On the basis of the confessions elicited during these terrible hours and the handful of supposedly damning artifacts from among his possessions, Vavilov was found guilty of espionage and anti-Soviet behavior. The punishment was decided on July 9, 1941. It was to be death by firing squad, along with the confiscation of all his possessions. Vavilov was fifty-six. He had two sons, a wife, and an ex-wife. He also had the largest collection of agriculturally important seeds in the world, the largest collection ever gathered, along with new crop varieties that those seeds had been used to breed. What would happen to all he had done? What would happen to his seeds? To his family? To the people in his employ?

Vavilov formally petitioned for a pardon. The pardon was denied. But in the days before he was to be sent to Butyrskaya prison to be executed, the Germans surrounded Leningrad, where Vavilov had stored most of his seeds and where he happened to be kept (little did his colleagues know). Moscow, too, was threatened. As a result, on October 24, Vavilov and the other prisoners were taken to eastern cities, to distance them from the war front. Vavilov was imprisoned in Saratov, Prison No. 1, on Astrakhan Street, in the same city in which he had begun his career as a professor.[6]

Vavilov lived on in prison, his will still present if diminished. He lived in a small basement cell with no window, a cell he shared with two other men. It had a light on all day, all night. It had one table and one bed. Two men slept on the bed, one on the table. They took turns. Vavilov gave lectures to the men about botany, they about their own fields. He also wrote a book, he would later tell a young woman whom he met in the prison, about "the history of worldwide agriculture." He had also finished a large work on plant breeding and plant diseases. These are likely to have been, assuming Vavilov's mind had not deteriorated, two of the most important books on agriculture and plants ever written. Somehow he had found a pencil and paper.

Meanwhile, Vavilov's wife, Elena Ivanovna Barulina, and their son, Yury, did not know where he was. Nor did Vavilov know where his family was. The extra tragedy in these dual mysteries, if any need be added to this story, is that by coincidence Vavilov's wife and son Yury had been evacuated from Moscow to Saratov, the same town in which he was imprisoned. They walked near the building in which he was housed.

On January 26, 1943, weakened by hardship and dysentery, Vavilov died of hunger and scurvy after more than a year of eating little other than kasha, salted fish, and more kasha.[7] The man who, in all of history, most eagerly fought for the diversity of crops died because of its absence.

Not long after Vavilov's death, his brother, Sergei, who had become an eminent scientist himself and had worked on Stalin's nuclear project, visited Leningrad. He saw Nikolai's apartment, which Stalin's men had, by then, long ago cleared out. A ballerina, Natalia Dudinskaya, was now living in it, practicing her dance steps on same floor where Vavilov had so often arranged his seeds, her feet arcing through the dust of their remains. Sergei also visited the institute, where the smells of human death and musty grain hung in the air. He looked around at his brother's seeds, still capable of germination. He felt raw and defeated and could not help but wonder why only seeds and not souls could be brought back from within those walls.

Despite the tragedy in Leningrad, despite the tragedy of Vavilov, his seeds made it through the siege. At war's end, almost all of most of the crop varieties Vavilov and his team had collected remained; many of them were replanted in the Soviet Union right after the war. In their growth, the seeds had the potential to offer a heroic conclusion to the tragedies, a rebirth and a memorial. But would they?

The immediate challenges were great. In the aftermath of the war, Soviet agricultural research, in ecologist and writer Gary Nabhan's words, ground to a halt. Most of the seed varieties Vavilov had brought back had not yet been tried in the field, nor would they be.

Breeding according to the insights of genetics and evolution stopped. Soviet textbooks were altered so as to reflect Lysenko's views. University courses, too. Not only did Vavilov die, so did work in the Soviet Union on modern genetics and evolutionary biology. As a result, Soviet agriculture actually moved backwards until the death of Stalin, in 1953. (Lysenko did not die until 1976, a reminder of the incredible recentness of this story.) Even after Stalin died, however, the recovery was not immediate. Soviet genetics, particularly as related to agriculture, lost three decades. In 1972, the Soviet Union bought four hundred million bushels of wheat from the United States, unable to supply its own people. The wheat was of a hard red winter variety originally grown in Russia. Some argue that the negative effect of Lysenko and Stalin on agriculture played the single largest role in the fall of Soviet Communism.[8]

In subsequent decades, however, thanks to dedicated scientists and staff, Vavilov's seeds, at least, survived. More than once, they just barely survived. They helped usher in advances in Russian agriculture, especially in the north. The northern distributions of corn, cotton, rice, soybeans, sorghum, tea, and pulses (beans and other legumes) in Russia have all increased, thanks to work based on Vavilov's collections. New varieties of crops were planted at research stations and then spread from farm to farm around particular regions. Other accomplishments were made possible thanks to the work of those who continued Vavilov's mission. More than four hundred of the varieties of crops that Vavilov collected or bred are now growing in fields across Russia. Four-fifths (which is to say nearly all) of the land in Russia is sown with seeds collected by Vavilov and/or his institute.[9] Their benefits vary by region but perhaps are greatest in those regions where crop yields were unpredictable from year to year, where families lived from hardship to hardship. It is estimated that thanks to the wheat varieties Vavilov brought to

Russia, the annual production of wheat increased by 80 percent relative to what it was at the time of his death. This is to say nothing of other crops.

But Vavilov's mission was global. And to some extent, having garnered seeds from around the world, the responsibility of Vavilov's institute was also global. Yet no one has provided a concrete estimate of Vavilov's global contribution. Those who save seeds on behalf of the future tend to be poor at measuring what they have done. They share selflessly. They "fall on their seeds" on behalf of people they don't know, assuming that someone else will do the same and that those seeds might be passed on not for just a few years but for a few millennia or more. They assume they are doing something for posterity. As a result, it is hard to sum up the value of such collections, hard to know what to count or when to stop counting. What, collectively, are they worth? What, even, is the value of Vavilov's collection to the Russian economy and people? How much is it worth in terms of continued funding each year? These estimates need to be made but haven't. Informally, those who gather seeds suggest that the answer is incalculably grand, worth dying for. But that very incalculableness has made it easy for governments to refrain from further investment, not only in Russia but also around the world.

Estimating the value of Vavilov's work is all the more difficult because to do so well would require telling the story of each seed that passed through Vavilov's hands and into the hands of others. For example, one variety of potato collected by Vavilov was taken to Hungary by Dr. István Sárvári. Sárvári used it to breed new varieties of potatoes resistant both to late blight and various viruses. His work paid off: some of the potatoes were resistant to both. Some of those resistant varieties were grown in Romania, where Scottish scientists encountered the potatoes in 1992. They negotiated a deal with the Sárvári family. Now the potatoes are being grown in the UK and elsewhere, sold as Sarpo potatoes. This entire chain of events was made possible by Vavilov's work and the work to save the potatoes during the siege of Leningrad. How many other such chains might there be?

But even the collections themselves (and the work of many thousands of farmers from around the world they build on) and their subsequent breeding are only part of what Vavilov contributed to the world. What lasts may be different from what was intended—in this case, what lasts may be inspiration. Vavilov's seed collection was not the first, but it was the largest and the first of its kind to explicitly attempt to conserve crop varieties from each of the major regions in which crops were domesticated and then to use them to breed new, useful forms. Many collections would follow this lead, sometimes directly. Vavilov persuaded the great botanist John Hawkes, for example, to take some of his potatoes back with him to England. Hawkes did. Hawkes, in turn, trained Carlos Ochoa, who went on to play a central role in the International Potato Center and its collections in Peru.

Where collections existed before Vavilov, his influence changed how they worked, their scope and grandeur. Abraham Lincoln established the United States Department of Agriculture (USDA) in 1862, in large part to collect seeds and distribute them among farmers. In 1898, the Office of Foreign Seed and Plant Introduction was added to the USDA; this office focused even more intently on getting seeds from around the world to farmers, shipping millions of packages a year. But it was not until after Vavilov's death that this collection was coupled with the establishment of large seed banks and experimental farms where seeds from those seed banks could be planted and crossed with other seeds. It was then that the National Center for Genetic Resources Preservation, including the National Small Grains Collection, was established in Fort Collins, Colorado. Vavilov's legacy is similar in other regions. His work led to the origination or expansion of nearly every other major seed collection in the world.

Among those directly influenced by Vavilov was Harry Harlan. While working for the US Department of Agriculture, Harlan traveled the world searching for new crop varieties, particularly varieties of barley, his specialty. It was during this search that Harlan and Vavilov became friends. Harlan helped Vavilov on his expeditions through North America and hosted him. The two sat in Harlan's living room and talked long into the night. On the floor or sitting in the corner during these conversations—stories of traveling the world, sagas about saving seeds—was Jack Harlan, Harry's son. Jack grew up breathing Vavilov's dream as if it were the ordinary sort of air everyone inhaled. So it was not surprising to Harry when, at the age of fifteen, Jack announced that he wanted to work with Vavilov. He planned to travel to Russia as soon as he could. It could have been a passing desire, but it was not. By the time he was in college, Jack Harlan was ever more dedicated to going to Russia. He even studied Russian at George Washington University while working on his bachelor of science degree to make such a visit possible. But when the time came, things had already gotten bad for Vavilov in Russia. Vavilov told Harlan's father, on his last visit to the United States, in 1932, that he would communicate in code whether it was safe for Jack to visit. If Vavilov began a letter, "My Dear Dr. Harlan," it was not safe to visit. If Vavilov began, "Dear Dr. Harlan," it was. Jack wrote a letter to Vavilov in the spring of 1937 asking to visit. He soon thereafter received a reply that began, "My Dear Dr. Harlan." Russia, it seemed, was not safe.

Jack Harlan did not go to Russia, but he did go on to a career studying the evolution of crop plants at the University of Illinois. Inspired by Vavilov, he collected and saved everything he could, even if it seemed useless. For instance, when visiting a wheat field in remote eastern Turkey, Harlan and his Turkish colleague, Osman Tosun, found "a miserable looking wheat, tall, thin-stemmed,"[10] and otherwise uninspiring. It broke in the wind. Worse yet, it seemed susceptible to leaf rust and lacked winter hardiness. And oh, by the way, it was also terrible for baking. But Harlan collected it anyway and entered it into his American collection of seeds—his germplasm collection.

In 1963, stripe rust was killing wheat across the northwestern United States. Breeders were searching for something resistant. They tried many different varieties, including PI 178383, the name given to the seeds from Harlan's trip to Turkey. Fortunately, the breeders found that PI 178383 was able to fight off stripe rust as well as many varieties of four other pathogens. PI 178383, that "miserable look-ing wheat," has now had its genes bred into the most common wheat grown in the Pacific Northwest of the United States. It saved farmers from many millions of dollars in crop losses. But the legacy of Jack Harlan would be something else, something more grand. Jack Har-lan was the one who reconsidered Vavilov's centers of origin, both to expand upon and improve them. Jack Harlan, too, would become, more than anyone else, the advocate for a particular part of Vavi-lov's vision. Vavilov was aware that crop varieties were being lost, but it was Jack Harlan who would sound the clarion call that saving the diversity of seeds was not just useful but also something that needed to be done immediately, before it was too late.

Today geographers and anthropologists often talk about the tra-ditional knowledge of peoples indigenous to particular places. Long before the words *traditional knowledge* were used, Vavilov under-stood the value of such knowledge. He learned many languages not because he could but because it seemed necessary in order to learn what various peoples knew about their crops and how to farm them.

The last twelve thousand years of agriculture have yielded an extraordinary diversity of crops. But the process of that flowering was slow and depended heavily on chance as a key element in the process. Mutation produced new versions of genes, and sex mixed those versions among offspring. Together, mutation and sex yielded variety, just as they had for the previous billion years of evolution. The role of farmers was to choose among the varieties, winnowing, favoring, and sharing favored forms. The modern synthesis offered a way to speed things up by strategically crossing crops rather than depending on chance. This new approach, Vavilov imagined, would move into the future, coupled with the movement of traditional vari-eties, the conservation of traditional knowledge, and the conserva-tion of the process of creating those traditional varieties. In other words, he would add to, not replace, the old ways.

It was in the United States, however, that attempts to use tradi-tional varieties of crops to breed new varieties most took hold, albeit in a particularly American way. Rather than try to produce new varieties that might grow anywhere, American scientists would come to focus on using pesticides, fertilizers, herbicides, and irrigation to make conditions as similar as possible everywhere and then to breed crops ideally suited to those conditions. This approach would come to depend disproportionately on one man, a man who both built on Vavilov's legacy and, as Jack Harlan would point out, set the stage for its destruction.

10

The Grass Eaters

We are in the midst of a mass extinction event in agriculture, at precisely a moment in history when diversity for further adaptation is most needed.

—Cary Fowler, in the preface to *The Heirloom Tomato* by Amy Goldman

Norman Borlaug grew up on a farm in northeastern Iowa near the town of Cresco. His mother, from rural Norway, worked hard. His father, from rural Norway, worked hard. Norman worked hard. Hard work came naturally to Norman, even if other things did not. He lived through the Depression and lived with its hard lessons. As a teenager, Borlaug decided to attend the University of Minnesota. But upon arrival at the university he failed part of his entrance exam and so spent much of the first year taking remedial classes at the so-called general college. Even then several of his advisers questioned whether he was ready for the four-year college. In the end, he was allowed into normal classes. He also wrestled. He was part of a very successful wrestling team with which he would come to travel around the Midwest. On that team, Borlaug was not the cleverest wrestler. Nor did he have the most natural talent. Nor was he the most athletic or the strongest. He was, however, even on a team with several national champions, the most persistent and the hardest worker. In his wrestling, in his classes, and in his life, he won by being tireless.

After finishing his undergraduate degree and his last year of wrestling, Borlaug went on to earn an MS in 1940 and a PhD in 1942, both in plant pathology. He worked with Elvin Charles Stakman, a man dedicated as no other to fighting (and defeating) wheat rust, and Herbert Kendall Hayes, a leader in breeding new varieties of wheat.[1] The work Borlaug did with Stakman and Hayes for his graduate degrees was fine but still somehow unremarkable. Yet it landed him a well-paying job with DuPont, working on new pesticides, antibiotics, fungicides, and preservatives. This was, for him and for his family, a story of success. But then World War II started. Like Vavilov, Borlaug tried to enlist in the armed forces. Like Vavilov, he was denied. His job at DuPont was too important to the war effort. He was to save the world one petrochemical at a time, whether those petrochemicals were to be used to control pests or in the production of new fabrics for the military, such as rayon.

Meanwhile, a crisis was brewing in Mexico. The yield of crops, including the key staples, corn and beans, were not keeping up with population growth. In addition, wheat varieties, first introduced to the region by conquistador Hernán Cortés, were besieged by one of the rusts Stakman studied, a stem rust (*Puccinia graminis tritici*). Wheat crops were destroyed by stem rust in 1939, then again in 1940, then again in 1941. Moreover, when the outbreaks of rust were bad in Mexico, the rust then spread north to the United States.[2] Stakman was called in to find a solution.[3] Stakman's first step was to enlist his former student, George Harrar, to run the project, which would focus on breeding new crop varieties. Next, he contacted Borlaug. In 1942, he asked Borlaug to leave DuPont and move to Mexico. Borlaug, although initially reluctant, ultimately agreed. The Rockefeller Foundation supported the overall project, which came to be called the Mexican Agricultural Program. It was one of many agricultural endeavors in those years made possible by American philanthropy.

Borlaug arrived in Mexico in October of 1944. There he was to work under George Harrar. Borlaug initially worked on corn, but his focus soon shifted to breeding a rust-resistant Mexican wheat. Once again, Borlaug was starting out in remedial class. He had never worked on wheat, nor had he ever left the United States, nor did he speak Spanish. His approach was blunt and forceful.[4]

Borlaug's work on wheat was based in the mid-elevation region of El Bajío in Central Mexico. In the traditional method of small-village farming, making a new variety of wheat relies on the chance or near-chance breeding of various varieties of the crop and the subsequent winnowing, one generation per year, of the best varieties for each individual soil or climate. For most of the history of agriculture, crops have been tailored to particular soils, particular climates, particular sets of pests and pathogens. Borlaug's plan was to take evolution into his own hands.[5] He would selectively perform many crosses, one variety at a time, until some offspring had the attributes that were desired. He would then breed those offspring with each other until they were all identical (or nearly so) and all reasonably resistant. The problem was that this took time. One made a cross, waited half a year, saw the result, and then, based on that result, started anew the next year. Borlaug was not going to wait half a year. He needed to figure out a way to speed things up.

He decided to start by going bigger. Rather than doing a handful of crosses—which is what Harrar had been doing before Borlaug arrived—with the wheat varieties already present in Mexico, he examined seeds in collections from all over the world, including every variety of wheat in the USDA vault. Based on those he deemed promising, he started to cross every possible combination of those seeds.[6] A by B, B by C, A by C, and so on. Some seeds came from Vavilov's collection. Others from Syria. Others from Japan. In the first year, Borlaug did more than two thousand different crosses. In the second year, he did even more. His hope was that some of those crosses would yield individual plants that were resistant to rust and grew fast. Borlaug wanted wheat varieties that would survive fungi, be sprayed for pests, and suck up whatever water and nutrition they could be given—crops that could be grown across millions of acres, regardless of where someone might plant them, so long as a farmer could get to them with water, fertilizer, pesticide, and a tractor.

He was asking a lot. Not surprisingly, of his nearly five thousand crosses, just two proved resistant. Neither had all the traits he wanted. He needed more variety. The resistant two, of course, might yield the answer when crossed with other varieties. But the odds were low. And then he had to wait another year to see if any of the next rounds of crosses, between the resistant varieties and more productive varieties, would work. Borlaug was frustrated, but he had an idea.

Borlaug knew there was another region in Mexico where wheat was being farmed, a region nearly a thousand miles north in the Yaqui valley of the Sonoran Desert, far enough away and low enough in elevation for the growing season to extend into fall and winter. The seasons of the Sonoran Desert are delayed relative to those in El Bajío, so in theory Borlaug could get the seeds from a first season of wheat growth in El Bajío (from May to October), plant them in irrigated fields in the Sonoran Desert, and get a second crop the same year (from November to April). This approach would be difficult. The roads to the desert ranged between terrible and nonexistent; he couldn't drive there directly. He would have to drive into the United States, to Arizona, and then back down to the desert, more than two thousand miles in total.

His plan also went against the basic rules of seed biology, which suggest that winter wheat and many other agricultural seeds need to go dormant for a while before germinating; they need a winter. It also went against the idea that seeds from a place where the day is long won't germinate in a place where the day is short, and vice versa. It went against decades of understanding. It went against the advice of Harrar—who predicted that the plan would fail and balked at the expense[7]—and Stakman. Add to this the fact that work conditions in the Sonoran Desert were unusually tough. Borlaug slept in a hayloft. He borrowed tractors and, in one particularly desperate moment, could be found pulling a plow through the soil with his own body in place of an animal. Borlaug was, in effect, a stubborn mule.

But it all worked. Borlaug was able to grow two generations of wheat in a year. He also managed to breed wheat varieties resistant to stem rust. As a bonus, the varieties, having been through the geographic gauntlet, grew if conditions were right regardless of how long the days were (in the lingo of scientists, they were "photoperiod insensitive"). They grew without having to go through a period of dormancy. They were tough and strong, and in the presence of fertilizer, irrigation, and pesticides, they grew nearly boundlessly.[8] By 1950, Borlaug had released eight new wheat varieties to Mexican farmers. By 1956, the seeds of these plants were planted across the country, and Mexico produced four times the harvest it had produced in 1945. Four times. Borlaug hadn't just created a new variety of wheat, he had also created a whole new approach to agriculture, one in which fast-growing plants, irrigated as much as they needed to be, fertilized and, in many cases, protected by pesticides, became the new model for a crop. The more important a crop, the more likely it would come to be bred using the Borlaug approach, now called shuttle breeding. He was speeding up the world.

Borlaug took his breeding approach global. Rather than just moving seeds between two sites in Mexico, he could shuttle them from country to country, getting half a dozen generations in a single year. This global shuttle breeding yielded varieties of wheat that were even more resistant to rust, including those now eaten around most family tables in the Western world. By some measures, Borlaug achieved as much change in wheat varieties in five years as had occurred in the previous thousands. But whereas the previous thousands of years had tended to generate diversity, Borlaug was producing simplicity, the single perfect form. And why not? There was one perfect way to build a spaceship; one perfect way to build an artificial heart. Surely the same was true of wheat.

Here was real success, but there was a problem. In a way it was the problem of too much success, though it was a little more complex than that. The wheat plants—which, as part of the new agriculture, were fertilized—grew too tall. Their stems could not support their seeds. They fell over, or lodged. Lodging had been seen before, when farmers applied heavy loads of manure or guano, but it was far worse in Borlaug's high-yielding varieties. The mystery was this: Why wouldn't the wheat plants, given more food, simply produce more seed? Why did they waste energy on stems when they didn't need them? Natural selection, after all, favors organisms best able to produce a large number of successful offspring.

To solve the mystery, one needs to remember that wheat plants did not evolve in a peaceful world; they evolved in the grasslands of the Fertile Crescent, where they competed with many other species of grasses and, most intensely, with individuals of their own species. Competition is always most intense within species, human territorial wars being just a particular manifestation of a general phenomenon. Wheat evolved, in other words, in fields in which much of its energy was spent waging war so as to be able to intercept the most sunlight and soil nutrients. In such a war, plants, even mothers and offspring, compete with each other for sun, water, and nutrients.

It is this competition that yields forests. Forests are what happens when trees compete with each other to get closest to the sun. The height of forests is determined by a balance between the availability of water and soil nutrients and the race to grow tall. In places where water and soil nutrients abound, such as in the parts of Northern California where redwoods grow, trees reach a hundred meters (three hundred feet) into the sky, wasting enormous quantities of energy trying to be slightly taller than their neighbors. The same is true of similar climates and soils in Australia, Japan, and New Zealand. We think of the tall trees in ancient forests as majestic; the truth is they are the macho a-holes of the plant world, fighting it out for sun when if they just agreed to share they could save all the energy they wasted on height and build trunks instead. Something very similar happens in wild grasslands, but grasslands tend to have frequent fires, so that the individuals that do best are those that grow tallest most quickly in the time between fires. Where fires are prevented by methods such as modern human management practices, forests quickly take over most grasslands.

Wheat evolved in a fire-prone grassland and so evolved genes that allowed it to grow fast and quickly produce seeds. Each wheat plant is trying to grow taller than other wheat plants, but we want something different from our farms. We want wheat that spends as little energy as possible on stems and as much as possible on seeds, on grains. This has always been a challenge in agriculture, one that has led traditional farmers to breed crops that are typically much shorter than their ancestors and have much larger grains. Wild cacao, for example, is a relatively tall tree, whereas domesticated cacao is something more like a tall shrub. Wild corn—teosinte—has tiny seeds, whereas domesticated corn has, well, corn kernels. But once Borlaug started to favor fast-growing crops, he was choosing crops that, in a given amount of time, grew taller, the opposite of what he wanted in terms of their stems. Worse, once those crops were fertilized and irrigated, they grew even faster, trying even more avidly to block the sun from other plants. Borlaug needed to stop his wheat from competing with itself so that it would produce as much seed as possible and as little stem. He searched the Agricultural Research Service's National Small Grains Collection, established by the USDA in Fort Collins, Colorado, for shorter wheat varieties but found none that seemed just right.

He didn't know that years earlier, on the other side of the world, a Japanese scientist, Gonjiro Inazuka (1897–1988), had begun working on just such a variety of wheat. Inazuka sought to develop a short variety of wheat for Japan. He worked at what is now called the Iwate Agricultural Research Center to breed one. The value to Japan of such a variety was the same as it would be to Borlaug in Mexico; it would be a wheat that would produce more seeds, less stem, and hence not fall over. For Japan, an island nation short on land and with a rapidly growing population, achieving more grain per hectare was a necessity. Another option to improve food security for the country was to colonize nearby regions, such as Formosa and Korea (which it did). A high yielding, short wheat variety, if it could be achieved, would not only provide sustenance but also help keep regional peace by feeding more people on the same amount of land.

In 1935, Inazuka bred such a variety, which came to be called Norin 10 (Norin was named for the Japanese Ministry of Agriculture and Forestry). The moment in which it was produced was a great one.[9] Then came World War II, slowing Inazuka's progress. In 1945, Americans dropped atomic bombs on Hiroshima and Nagasaki. Japan surrendered on August 15. General Douglas MacArthur entered Tokyo, and the Americans took over the country from 1945 to 1951.

MacArthur was charged with leading the reconstruction of Japan. It was a reconstruction that, in many respects, remodeled

Japan on American ideals; this was particularly true for Japan's agricultural system. MacArthur established a Natural Resources Section in his army of occupation that was charged with solving food shortages after the war through innovation in agriculture, which, superficially, meant the transfer of American agricultural methods and technology to Japan. But the Natural Resources Section also allowed Americans to benefit from Japanese agricultural successes. Here, wheat is exemplary.

The Natural Resources section thought that Japan needed help with wheat. In December of 1945 a man named Cecil Salmon was called in from his desk job at the USDA in Washington, DC, and charged with helping improve Japanese wheat. Salmon, like Borlaug, was a student of the University of Minnesota, where he was also influenced by the vision of E. C. Stakman, and he was happy to get out of the office and back into the field. While in Japan, Salmon helped to set up a national research network for wheat. This required him to travel from field to field around Japan. As he did so, he found as many opportunities to improve US agriculture based on Japanese successes as the reverse. The Japanese had bred many varieties of crops with traits that were missing from the varieties being worked on by American crop breeders. One of these was Inazuka's short wheat, Norin 10. When Salmon left Japan for the United States in the summer of 1946, he brought Norin 10, which he encountered at a research station in Honshu, and several other wheat varieties with him to Washington, DC. From there the wheat traveled to Orville Vogel at Washington State College; Salmon thought Vogel might be interested in the variety. Vogel recognized the value of the seeds and crossed them with two other wheat varieties. Borlaug requested samples of the seeds and, soon enough, samples of both of those crosses arrived in the mail in Mexico.

For Borlaug, the short Japanese wheat was a kind of miracle. He bred it with his high-yielding wheat, and it produced a high-yielding, rust-resistant short wheat that even when well watered and fertilized did not grow too tall. It also produced more seeds, more grain. With it, yields doubled again. Thanks to Borlaug, Harrar, Stakman, and Vogel (and, unwittingly, Inazuka), in 1965 Mexico increased its wheat yield *tenfold* relative to what it produced in 1945. These men had changed Mexico. Next they would change the world.

For context, it is useful to compare the yield of wheat in Mexico in 1965 to historic yields. Wheat grown in the wild, like the long wheat gathered in the Fertile Crescent, would have yielded, at most, perhaps a tenth of a ton of wheat per acre. By the tenth millennium BCE, once wheat was domesticated, it would produce a half ton per acre. Over the following twelve thousand years of agriculture—hard years, years in which science helped farming—the yield pushed up to two tons per acre. By 1965, Borlaug and the global wheat-breeding community had pushed this two tons up to six, roughly sixty times the yield of wild wheat. For his part in this work, Borlaug would be awarded the Nobel Peace Prize in 1970.

· · ·

As the years passed, Borlaug became director of the wheat program at CIMMYT and began to work in India, then Pakistan. By that point, Borlaug had become, as the scholar John Perkins has written, a crusader "who took the word about higher yields to anyone who would listen."[10] India and Pakistan did. With this word, with this approach, came many associated changes. Borlaug's end product was not just a new variety of wheat; it was also a new approach to agriculture. Borlaug, like his adviser Stakman and his colleagues, was globally engaged, focused on results, and friendly to both government and industry. Each seed embodied this perspective; each seed was the result of focused work, done internationally, involving a partnership with—really, a need for—industry. Tractors were needed to deal with farming on a large scale. Gas was needed to fuel the tractors. Pesticides, fertilizers, herbicides, and other chemicals (themselves nearly all produced from petroleum) were needed, too. Roads and railroads were needed. These things all went with the new agriculture. As did higher prices of land and, in many cases, the concentration of farms in fewer hands. The new agriculture changed farming, which changed how people lived, which changed where people lived. The new agriculture began to shift families away from small towns and into cities. The new agriculture precipitated the largest social change since the origin of agriculture itself. It is hard to overestimate the impact that Borlaug, Harrar, Stakman, and other crop scientists of their generation had on agriculture and, as a result, the average daily life. They shaped where we live, how we live, and even, at the global scale, how many of us have lived.

While Borlaug continued to work on wheat, spreading the gospel of high yields to places such as India and Pakistan, Harrar set about revolutionizing rice, working through the International Rice Research Institute, in the Philippines. This endeavor, too, was funded by the Rockefeller and Ford Foundations in 1959, to the tune of $7 million (roughly $50 million in 2016 dollars).[11] Harrar got started by writing to sixty countries to see who might share rice seeds. He was essentially outsourcing Vavilov's approach, relying on the postal service rather than his own peregrinations. It worked. Many scientists in many regions sent seeds. Harrar was able to build on the model he and Borlaug had used for wheat in Mexico to rapidly breed the seeds they received, but the starting place was modest. Whereas Borlaug was working with hundreds and then thousands of wheat varieties, Harrar started with just a handful of rice seeds—the handful he was sent, enough to do a very modest thirty-eight crosses. Amazingly, out of these thirty-eight crosses he found a variety that was short enough to avoid falling over, resistant to the primary pathogens, and high-yielding.

Each of the crops Borlaug, Harrar, and Stakman worked on, each grass, changed in those years, fundamentally, in terms of how it was grown. These changes and the changes they brought about in the world would come to be called the Green Revolution, where "green" refers only to the amount of food being produced and not to the sustainability of the approach. During the Green Revolution, a few varieties of wheat and rice took over in field after field. Similar changes would follow in other crops.

As they spread, the Green Revolution varieties not only shifted what was farmed, they also shifted what we ate. They shifted it in countries with enough resources to pay for the necessary chemicals and enough political stability to invest in roads and rail lines. In addition, they shifted it in countries that the US government worried might tip into Communism. They were all northern countries or the temperate parts of tropical countries (e.g., India and Mexico). In these places, crops yielded more and so became cheaper. In 1839, the United States produced just shy of four hundred million bushels of grain. In 1958, it produced a thousand times more.[12] A thousand times! Meanwhile, the population had increased just tenfold. Diets changed accordingly. As of 2016, wheat accounts for 20 percent of all food calories consumed on earth. It is important not only in those regions where we picture "amber waves of grain"—North America, Russia, and Europe—but also in many poorer regions, places such as Uganda. Think of a plate of food, the average global meal, and divide the part of it derived from plants into fourths. Approximately one of those four parts consists of wheat and one consists of rice, which is to say that nearly half our global diet is composed of just two of the grasses bred by Borlaug and his colleagues—which is to say nothing of corn and sugarcane, both of which are also grasses.

Several things were inevitable as the crops bred by Borlaug, Harrar, and their colleagues spread around the world. One was that the productivity of fields increased. Another was that the size of populations in these regions also swelled, much as they had swelled in earlier generations, when the age of discovery first moved crops around the world and those crops escaped from their enemies. Another was that the pollution associated with soil erosion, fertilizers, and pesticides increased exponentially. Pollution is the waste of human living that is not reused. Every agricultural system has produced pollution. But the new Green Revolution systems produced pollution that acted in far different ways. Fertilizers from fields fed not only crops but also algae in rivers, lakes, and ultimately oceans. Pesticides killed not only the pests that were eating crops but also the insects that ate the pests and, when pesticides were applied too generally, other animals. DDT weakened the eggshells of birds. It also afflicted humans. Herbicides posed similar problems. All this followed like a dust cloud behind the new agriculture of the Green Revolution wherever it went. But there was also something else.

The Green Revolution was economic. It was economic in its consequences (many hundreds of millions of people, at least, benefited).

It was economic, too, in that it turned agriculture from a local to a global economic activity. This had happened before—in ancient Rome, for instance—but the new geographic scale of the markets was larger than it had ever been, global or nearly so. And once a farm started to use the crops of the Green Revolution, it was also wed to use fertilizers, and in a way that bound the future to the same model. Once a farm started to use the crops of the Green Revolution, it had to use fertilizers and, in most places, irrigation, pesticides, and herbicides. These compounds created an environment into which Green Revolution crops could be planted. It created what was, in effect, a new biome, one in which the outside world (climate, seasons, and soils) influenced agriculture only in the most extreme cases. Some places are too hot or too cold for a particular crop in this new world. But most of the variations in rainfall and pestilence are removed through chemistry and plumbing. It was a new world in which any crop that could be made to generate high yields by crossing it with another plant (or, later, through engineering) could be produced in great quantities, and the subset that could be made to produce the very most was chosen—by farmers, by the market, by you and me when we go to the grocery store and buy the cheapest box of anything—to continue producing.

The new crops, in short, created a new ecological and economic realm embedded in an American-style agrarian capitalism.[13] The economic model associated with the new agriculture was one in which an already wealthy entity could make more money and produce more food, but it was also one in which the farmer was tied to a model that required funds for commercial seeds, commercial equipment, and commercial fertilizers and pesticides. This need would only increase. The farmers who could buy and use the new technologies did better at the expense of those who couldn't (and who, out of need, often moved into other professions). Most land came to be farmed based on seeds from afar, and this, coupled with the smaller number of farmers, meant that traditional, locally adapted varieties of seeds began to disappear. Inasmuch as seeds were being sold for this new ecological realm, Borlaugia, it meant that there were companies making a great deal of money off of farmers, companies that controlled the key elements of these farms. The seeds that were once swapped from hand to hand were sold for an arm and a leg.

Then there was the question of how long this new Borlaugian world could exist without disruption. As Borlaug was quick to point out, the crops he helped to breed would not do well forever. He gave them thirty years until the pests and pathogens figured out a way to deal with the resistance he had bred into crops or a way to deal with the pesticides being sprayed upon them. If he was right, this meant that well before the thirty years elapsed (roughly in the early 1990s), someone would need to breed a next generation of crops and engineer a next generation of chemicals. But one thing would be different: that person would have to do so by relying even more on seed collections and less on traditional knowledge, which had, by then, started to disappear. That person would have to hope that someone had collected the right seeds and saved them, though that would not prove to be quite enough. Someone would also have needed to save wild nature — the microbes, insects, and trees.

11

Henry Ford's Jungle

Like a swarm of annihilating grasshoppers, the inhuman
gang of rubber barons continue to press forward.
> —Dr. Theodor Koch-Grünberg, quoted in
> *The Domestication of the Rubber Tree*
> by Richard Evans Schultes

Nature foils and surprises. No one—no country or culture, no
person—is immune to its laws. Not even Henry Ford.

Henry Ford's legacy is part of our modern agricultural story in
many ways. He turned cities into suburbs, sent us sprawling out on
vast networks of roads, roads across America that shaped where and
how we grew food. Ford's cars and tractors helped make the Green
Revolution possible. But the relationship is mutual. Cars also depend
on agriculture; they require rubber, and rubber grows in trees.

Cars require oil and steel. Wars would be fought over oil; steel,
too. Cars, though, also require rubber. Rubber should have been the
easy part. It grows in the tree *Hevea brasiliensis,* native to parts of
the rain forest south of the Amazon River. The rubber that comes
from *Hevea brasiliensis* is the tree's defense against herbivores. Bite
a leaf or stem, and the rubber will ooze out and coat your mouth
with a gluey toxin that flows just beneath the bark's surface in wide
veins. In many Amazonian cultures this goo was used to make
waterproof objects, bags and balls, but the tree was more important
for its food. The seeds can be eaten.[1] For the first few hundred years
of colonial rule Europeans ignored the seeds and saw no value in the

rubber. Then in 1770, an enterprising British chemist, Joseph Priestley (who would also go on to discover oxygen), noticed that he could erase his pencil scribbles with the latex once it was coagulated, which is to say, turned into rubber. This created some demand. In Brazil some people made shoes and other items out of the rubber, but they were of poor quality. They were brittle: they broke in half, or if it was really hot, the shoes melted onto the city streets.

Two inventions changed everything. The first was the raincoat. Charles Macintosh figured out a way to make cloth, and eventually coats, waterproof by layering fabric sandwich-style on either side of a piece of thin rubber (hence the word *mackintosh* as a synonym for "raincoat"). The second was vulcanization. Vulcan is the Roman god of fire, the fire of invention and hard work, the blacksmith's flame. Vulcanization was invented in 1839 by Charles Goodyear, a poor man with rich dreams. He spent years of his life trying and failing to make rubber more durable. One day he mixed rubber and sulfur and put the concoction on the stove. It boiled and spat. He pulled it off and let it cool and then noticed that the rubber was no longer sticky. Yet it was still strong and bendable. The sulfur, Goodyear would come to realize, added elasticity and resistance to the rubber, making it more stable. With vulcanization, rubber could be used to make many products, including, eventually, tires.

The rubber market boomed, and the boom sent boats down each of the main tributaries of the Amazon—the Xingu, the Tapajós, and the Madeira—in search of more trees and people to tap them. The tapping process began with a diagonal cut made toward the bottom of the tree, from which the latex wept down into another cut, this one vertical. The vertical cut led to a cup or bucket into which the latex poured. Rubber tapping started early in the morning, and the tappers would go in a circuit quickly, covering hundreds of trees before the temperatures climbed and slowed down the flow of latex. With time, each cut in the rubber tree would heal over. When it did, another, higher cut was made. In many places in the Amazon today one can find trees with cuts extending twenty feet up, a measure of the hard work of the tappers. Those who did not work hard enough had their hands or fingers chopped off.

For years the demand for rubber increased. The price of rubber rose tenfold in the decade leading up to 1870 alone, and with it the horrors perpetrated on Amazonians, until in much of the Amazon nearly every tree was being tapped. The Europeans and Americans, meanwhile, got rich on rubber.

Everything would soon change. In 1876, the Royal Botanic Gardens, Kew, commissioned Henry Wickham, an explorer and entrepreneur for whom success had proved elusive, to gather the seeds of rubber trees in the hopes that those seeds might be planted in the British colonies in Asia. Wickham gathered seventy thousand rubber tree seeds from a single site in Brazil—Boïm, where the Tapajós River flows into the Amazon River.[2] He sent them via steamship to Liverpool. From Liverpool the seedlings traveled by train to Kew, where the director (and close colleague of Charles Darwin), Joseph Hooker, anxiously awaited their arrival.[3] When the shipment arrived, seven thousand of the small plants were still alive. The staff at Kew continued to grow the seedlings. Of the seven thousand, twenty-eight hundred were in good enough condition to be sent on a barge down the Thames to be loaded onto a British India liner.

Once the seedlings arrived in Asia, eager farmers planted the subset that had survived densely, so that the trees would grow up close together, side by side. They were planted in the Dutch East Indies (now Indonesia) and Malaysia on land where coffee rust had destroyed much of Asia's coffee production.[4] Planting trees in monocultures had led to coffee's demise in the region. But no one seemed to be paying attention to history.

The trees grew dense and profitable, much more so than the plantations that had been attempted in the Amazon.[5] Initially, few gave much thought to why this might be. It was a sort of miracle. Production of tires—and, by extension, cars—boomed. Seeds were later distributed throughout tropical Asia to former Dutch and British colonies (Queen Victoria would eventually knight Wickham). Throughout tropical Asia forests of rubber now grow as a result, forests with trunks so close that their canopies touch and shade the ground.

From Wickham's original trees, seeds were chosen that grew into trees that produced even more latex. In natural forests, the latex of rubber trees eventually stops flowing once the latex, like blood, begins to coagulate. Isoprene droplets fuse together and clog the specialized cells, lactifers, through which latex flows. But seeds were chosen each generation from trees on the Asian plantations that were increasingly unlikely to clog and more likely to bleed white, their veins perpetually open. Soon 90 percent of all the rubber in the world was being produced in tropical Asia (not far, fatefully, from Japan), at far greater volume per acre than anyone in the Amazon ever imagined possible. In 1912, 8,500 tons of latex were being produced in Asia. In 1914, 71,000 tons. In 1921, 370,000 tons.[6] By the time Henry Ford started making cars with rubber tires, essentially all the rubber in the world was coming from Asia, from the descendants of a tiny subset of Wickham's seeds.

. . .

Henry Ford didn't like being dependent on tropical Asia in order to make cars. He wanted to control the chain of production. To achieve that, he needed to grow his own rubber. He decided to create a giant plantation, one unlike any that had ever existed. It would be a utopian plantation where the workers lived in harmony, ate good food, refrained from sin, and were healthy, free of jungle parasites and pathogens.

Ford ordered this new civilization, which came to be called Fordlandia, to be hacked out of the rain forest in Brazil. The area cleared was one million hectares in size. It took hours to drive from one edge to the other. He had hundreds of houses built and hired more than two thousand men to plant, from seeds, two hundred thousand trees. Anyone could be hired to do anything, Ford knew, and so he could not conceive why his plantation would be any different. He wanted Fordlandia to mirror his Michigan assembly lines, which were described in 1914 by the journalist Julian Street:

Fancy a jungle of wheels and belts and weird iron forms—of men, machinery and movement—add to it every kind of sound you can imagine: the sound of a million squirrels chirking, a million monkeys quarreling, a million lions roaring, a million pigs dying, a million elephants smashing through a forest of sheet iron, a million boys whistling on their fingers, a million others coughing with the whooping cough, a million sinners groaning as they are dragged to hell—imagine all of this happening at the very edge of Niagara Falls, with the everlasting roar of the cataract as a perpetual background, and you may acquire a vague conception of that place.[7]

.   .   .

The assembly lines were as riotous as a jungle. Now Ford wanted to turn an actual jungle into an assembly line, one that began with soil and ended with rubber, one where all the machinery he really needed was trees. To do this, Ford imagined that the chief task was not reinventing the machine, the trees, which could just be planted, but instead reinventing the people, the society responsible for the work. Ford's two thousand men lived on the plantation in houses and barracks; the plantation's population would ultimately rise to twelve thousand. In order to ensure the health of their community, the men were forbidden to drink or smoke, and those who were not married were forbidden to bring women to their houses. They were fed for free, but their food came from the midwestern United States rather than Brazil—wheat and potatoes rather than corn and beans. Diversion could be found in churches, a recreation building, a golf course, and a library. Ford, who never visited the site, seems not to have considered that none of the men played golf, that most were only nominally Christian, and that many could not read. He seems not to have considered that not all men were the same as he was. As Greg Grandin notes in his book *Fordlandia,* "With a surety of purpose and incuriosity about the world that seems all too familiar, Ford deliberately rejected expert advice and set out to turn the Amazon into the Midwest of his imagination."

Some aspects of Ford's plantation were visionary—just as they were in Michigan, where a great deal of what was challenging in the auto plants (the monotonous work, the terrible noise, the banging tedium) was made palatable by a dependable living wage. Similarly,

Ford paid relatively well in the Amazon. He also established a new kind of tropical medicine, and both malaria and parasitic worms were largely eradicated from Ford's plantation, even though the same could not be said of almost anywhere else in tropical Brazil— or the tropical world in general—at the time. This achievement depended on good medical care as well as investment in public health, including, for example, a water filtration system that provided, daily, half a million gallons of clean, chlorinated water to the plantation. Then there were the trees.

Disregarding the conditions and the tendency of rain forests to thwart the ambitious delusions of white men, Ford estimated that, once mature, his plantation's trees would yield enough rubber for the tires of two million cars. At first they seemed to be growing flawlessly. The trees produced latex in the first few years. It bubbled from them. And why not? Rubber trees grew on Asian plantations. It was the growth of the rubber trees and the prosperity they implied, along with the healthfulness of Fordlandia, that kept people there. Healthy families, healthy trees. Also, this healthy life was made more tolerable for the workers thanks to the emergence of a cluster of businesses that opened on an island upriver that provided those recreations unavailable in Ford's utopia: Brazilian food, Brazilian drink, cigarettes, and Brazilian women. Compared to neighboring communities in Brazil, Fordlandia seemed almost too good to be true.

Ford did not know enough about rubber trees to know what to worry about. Nor had he sent a botanist, a plant pathologist, or an entomologist to the Amazon. He sent engineers and businessmen. As a result Ford did not understand his most likely foes. But many scientists did. In the Amazon, they knew, rubber trees contend with parasites that the trees planted in Asia escape. Among them is a species called South American leaf blight (*Pseudocercospora ulei*), an ascomycete fungus.[8] The rubber trees invest heavily in repelling and resisting insects, but because of history and chance, they invest little in avoiding fungi. The men working the trees, though, had heard about the blight. It was a primordial monster, a jungle demon that lurked in the deep forest, beyond the power of ordinary explanation. It was a monster, they believed, to be feared.[9]

Ford did not fear the jungle; he did not fear biology. His senior staff reported to him that in the forests of the Amazon the rubber trees looked healthy. Ford's trees looked healthy, too. What Ford was missing, and what anyone who had previously tried to establish a rubber plantation in the Amazon (and there were many) could have told him, was that leaf blight, the worst pathogen affecting rubber trees, does not strike until trees are mature, and it strikes worst where trees are dense.

. . .

Many rain-forest trees, including cacao, mangos, avocados, and coffee, have evolved fruits that attract vertebrates who carry them elsewhere in the forest. Trees, it turns out, tend to be bad mothers, or at least bad mothers to live near. Seeds that fall right beneath their mothers compete with them for sun. They also stand a high likelihood of catching whatever pathogens their mothers have. Stay by Mom and risk getting her disease. Stay by Mom and risk suffering a dark and pestilential life. As a result, species in which mother trees drop their seeds to the ground unaided have tended to go extinct. Those, on the other hand, that give their seeds wings so they might fly or fruit so they might catch a ride have prospered. The advantages of escaping one's mother seem especially pronounced in the tropics, where pathogens and pests are exceptionally diverse.

The fruits of the rubber tree have evolved what biologists call ballistic dispersal. As the rubber tree fruit dries out, it twists, explosively; during the explosion its seeds are flung as far as a hundred meters (328 feet). Many seeds land in rivers and are carried even farther. Tens of kilometers. Dozens of miles. This is far enough for the seeds to survive and avoid infection at least once out of every thousand attempts. But on his plantation, Ford was working against the grain of nature. He was doing exactly what mother rubber trees fight so hard to avoid. He was planting seeds if not below their mothers then right next to their siblings. The effect was the same. There was a high chance that if the tree next to you had a pathogen,

you would, too. This is what had happened in other rubber plantations in South America. In Suriname, forty thousand trees were
planted in 1911. By 1918, leaf blight arrived, spread from tree to tree
to tree, and all forty thousand trees had to be cleared. If Ford knew
about this history, he ignored it. And anyway, his trees continued to
look healthy as late as 1934. The canopies of the trees were so full
that they almost touched each other. Somehow Ford was defying the
ancient tendencies of nature; perhaps he was just that powerful.

He was not. Precisely when the canopies began to touch each
other, leaf blight arrived. In 1935 it showed up at one edge of the
giant plantation. At first it began to afflict the most mature trees,
the ones producing the most latex. It turned the leaves pocked,
green-black, then rotten, whereupon they fell to the ground, leaving
the trees naked, as though the plantation were a Michigan forest in
the fall. The trees would try to grow again, but the shoots would
stunt, able to produce only small leaves that also blackened and
withered. From the old trees, the blight spread to the younger trees,
then even to the nurseries of small trees and saplings. Within the
year, the blight had defoliated every tall tree and most of the others,
too. But the trees were of several varieties, including those repatriated from Asia, those especially good at producing latex. For a
moment, the hope as the blight spread was that the repatriated Asian
trees would be spared. This was a naive hope, and it was in vain.
They were not resistant. In fact the most remarkable thing about
them was just how quickly they lost their leaves and then, without a
spot of green, just how quickly they died. Henry Ford's industrial
model worked inside a factory. It worked in realms where nature
was not a factor and men could be controlled. It did not work in
nature; nature obeys rules, but they are different from those of the
assembly line.

But nobody messes with Henry Ford, and so in 1936, just a year
later, Ford had his enterprise moved to another site, where he once
more had a hunk of rain forest cut down, an even bigger hunk; it was
near to the town where Wickham had first gathered his seeds. Ford
called the new plantation Belterra. It was flatter and less susceptible
to morning mists than was Fordlandia, which seemed to encourage

the blight. The soils were better, too. In Belterra, Ford's team built a city modeled even more directly on a Michigan town. The Cape Cod houses matched those Ford had built in northern Michigan. The hospital bore similarities to the Henry Ford Hospital in Michigan and was, in some respects, even more modern. Then the trees were planted in even greater numbers—five million seedlings and seven hundred thousand trees. This time the trees grew even better. Straighter. Faster. With more latex. Then disaster arrived. It was not the blight (not yet) but rather a plague of insects and other pests, pests that ate the trees from one end of the plantation to the other. Lace bugs. Red mites. Whiteflies. Small black ants. White weevils. Leafhoppers. Treehoppers. Moths. Thousands of people were tasked with picking off the pests by hand. New pesticides were developed, based on fish poisons used by the workers. Then a molelike animal chomped at the roots. It was as if some jungle god had cursed the trees and called in the attack. The trees lost their leaves again and again. Were replanted. Were tended to, triaged, and protected. In 1937, James Weir, one of the Americans working at the site, was charged with finding seeds of other varieties of rubber, varieties able to deal with the pests or with the blight if it came back. Weir pretended to look for seeds but instead fled back to the United States, to Cape Cod. Then the blight reappeared. Once more the leaves disappeared from the trees, and this time they did not grow back.

Ford was done. He'd lost millions of dollars, to no avail.[10] He was able to grow neither a utopia nor a plantation. Nature won. The jungle was not just another assembly line, a reality we are still coming to terms with. As a result, Ford and the rest of the auto industry were, at the end of it all, as dependent on Malaysia, Singapore, and Asia in general for rubber as they had been when Ford first started cutting trees. Just a few years later, in 1942, as Hitler's troops were surrounding Leningrad and Norman Borlaug was traveling to Mexico to start breeding wheat, Japan cut off the supply of Asian rubber to the United States. If daily life and, perhaps more important, the war were going to continue, the United States desperately needed more rubber. Desperately. No new supply was forthcoming. Belterra was shut down, and no other plantations of any scale existed in the

Americas. The president ordered rubber conservation. Those few places that produced any rubber at all were squeezed. Fortunately the US government had just enough foresight to prevent absolute pandemonium. It funded attempts to create synthetic rubber, which had always been the dream of Thomas Edison, Ford's longtime friend.[11] The government also hired biologists to look for more wild rubber near the Tapajós and along each of the other tributaries of the Amazon—the way people had done before Wickham took seeds to Asia, the way they did in places where the inhabitants were poor enough for the low price of rubber to get them back on the footpaths.

Engineers worked frantically to produce synthetic rubber that was cheap enough to be mass-produced and strong enough to be used in tires, which at the time were made entirely of natural rubber, both on the surface that contacted the ground and the tires' sides. Not only was synthetic rubber produced, an entire industry also sprang up in order to produce it in huge quantities, enough to put tires on every plane, truck, and car needed in the war. Here was a victory for technology! We often seem to imagine that technology can do this, that it can produce a solution in the nick of time. Humanity is, it is said, infinitely ingenious. Synthetic rubber was the proof. But this ingenuity knows limits, though they would not be felt immediately in the case of rubber. The most immediate consequence was that synthetic rubber helped win the war.[12] In fact without it, the war would have been lost. Before the war, basically all rubber used in the United States was natural rubber grown in Asia. But after the war, synthetic rubber became the mainstay of industry, and by 1945, more than 90 percent of the rubber used in the United States was synthetic rubber, made in America.

. . .

Natural rubber, derived from plant latex, is complex stuff. It is an emulsion of large molecules composed of carbon and hydrogen. The building blocks of these large molecules are units of isoprene. In *H. brasiliensis,* the individual molecules contain thousands of isoprene units chained together, a great chain of carbon and hydrogen called polyisoprene. The length and complexity of these chains make natural rubber strong. Vulcanized rubber is even stronger in part because vulcanization actually links multiple molecules of polyisoprene together in even longer chains. To break vulcanized rubber apart, one must break apart the polyisoprene. Synthetic rubbers, on the other hand, have far shorter molecules, which makes the products they comprise more brittle and easier to break.

It should, it seems, be easy to make a synthetic rubber that matches the natural polyisoprenes in their size and complexity. It is not. As a result, synthetic rubber is neither as strong nor as long-lasting as natural rubber. Tires made of synthetic rubber wear out quickly. One can drive to California and back on natural rubber tires. With synthetic tires, one would be lucky to get to Colorado. Tires made of synthetic rubber are also too brittle to be used safely on airplanes (though during World War II, they were anyway). Yet the success of early synthetic rubber suggested that even greater successes were on the horizon. Surely synthetic rubbers that truly matched natural rubber in quality and utility could be invented. It wasn't to be.

In part the problem was the oil crisis brought on by the embargo against the United States by oil-producing countries in 1973. Natural rubber requires some oil to produce in terms of shipping and processing, but synthetic rubber requires much more. As a result, when the oil embargo hit, the price of synthetic rubber went up drastically, as did its price relative to natural rubber. When it did, the use of natural rubber increased. The use of natural rubber might have been reduced at the end of the oil embargo, but it wasn't. The reason was something entirely unanticipated—radial tires.

In the earliest rubber "bias-ply" car tires, cords of rubber were laid down like braids, running roughly parallel to the direction of the tire's movement. In radial tires, the cords were laid down perpendicular to the tire's movement. Radial tires were invented in the early 1900s but did not become popular among car manufacturers or consumers in the United States until the late 1960s. Eventually they took over the market, and now only antique cars are made with the

braided bias-ply tires. All this would be one more detail of the history of cars—a sort of boring detail, at that—if not for a key feature of radial tires: they require natural rubber in their sidewalls in
order to be strong enough to be used on cars. If you go out and kick one
of your car tires right now, the part your toe hits will have come out of
a tree grown in Asia. As a result, with the switch to radial tires and the
continuing increase in the demand for cars, the consumption of natural rubber increased dramatically in the 1970s. It has increased
every year since then, for the last four decades. The demand for natural rubber in 2016 was twelve times what it was in 1940. It is
expected to double by 2025. And the rubber is still nearly all from a
single variety of rubber tree grown in tropical Asia.

Fortunately, even after Ford's failed project researchers continued to try to find or breed varieties of rubber more resistant to leaf
blight. In the years after World War II, Richard Evans Schultes was
the key figure in these efforts. Schultes was initially sent to harvest
rubber in the Amazon during World War II. He gathered plants on
his expeditions that might have immediate use, but he also gathered
many more that did not have any obvious value at the time—those
he thought might be of use in the future. Assisted by indigenous
peoples and their traditional knowledge, he found many different
varieties of *Hevea brasiliensis:* some tall, others short; some with
lots of latex, others with little; and, most important, some with no
resistance to leaf blight and a few with complete resistance. He also
collected seeds of nine species of wild *Hevea* other than *H. brasiliensis,* two of which appeared to be able to produce latex that could
be used for rubber.[13] The other seven, all of which produce latex
that is hard to coagulate or in other ways suboptimal, might still be
useful when bred with varieties that are strong latex producers. He
gathered these species and their varieties from forests that were
either conserved or too remote to have been cut for timber or agriculture. He collected them with the wild obsessiveness of a man who
thought he was saving civilization. He collected millions of seeds. At
great personal risk, Schultes then took these seeds over mountains,
down and up rivers, on boats, and on planes back to those who

could use them in breeding programs. The breeding programs were scattered throughout the Americas, but the most ambitious effort was in Turrialba, Costa Rica.

In Turrialba,[14] the seeds Schultes collected were grown and fiddled with by Ernie Imle, among the very best breeders of tropical plants in the world. If Schultes was the gathering adventurer, Imle was the monk in the garden, slowly, meticulously bending nature's diversity to society's needs. Over the course of a decade, Imle and his team planted the varieties brought to them by Schultes. They then bred and rebred the plants. Breeding trees is hard work. Imle had to be creative. He had a sense of urgency about the work before him. His long-term goal was to breed varieties that were fully resistant to leaf blight and produced large quantities of latex. In the short term, though, he had an alternative plan. Imle was working to graft varieties of rubber that had coppices resistant to leaf blight (since blight only attacked the leaves) and stems that produced large quantities of free-flowing latex. Grafting trees together was cheap, and if he found the right combination, it might just work.

By 1951, Imle, as reported by Wade Davis, a student of Schultes's, had produced grafted trees in which the latex flowed as well and for which the coppices were resistant to blight. This was a major breakthrough. Imle just needed a few more years to get everything right. Such grafted trees could be planted in Asia preventively, but they could also be grown productively in the regions in which the leaf blight was already present. Schultes and Imle had carried out perhaps the most successful and systematic attempt to produce resistant, productive trees in history. Imle wasn't done, but he had achieved a major milestone. To make this all happen, Imle was necessary. Schultes was, too. As was the knowledge of the indigenous peoples with whom Schultes interacted. As were the rain forests to which Schultes traveled.

If this were fiction, this is where I would tell you that the plants Imle bred are now being used to start new plantations of rubber resistant to leaf blight, plantations that reduce the extent to which the world's rubber supply is in danger. But it isn't fiction: it is the very real story of science, nature, agriculture, and politics, and in that story the conclusion embodies disaster as often as it does success. In 1953, because of political fighting in Washington, DC, and a focus on putting out what seemed to be more urgent fires, the budget for the US rubber program was cut entirely.[15] The program at Turrialba was shut down. What's more, instructions were issued that the materials at the site should be removed or destroyed. Employees from the State Department took away notes and data from the breeding program. In other cases of threats to the plants on which we all depend, someone has saved the plants in the nick of time. But no car was waiting to save the rubber seeds from Turrialba, not even the trees. The seeds went missing, as did the files concerning them. The trees themselves were cut down.

As a result, Schultes's collections sit largely ignored. Imle's new varieties are gone, and we are more at risk for the loss of rubber today than we were during World War II. Leaf blight poses as much of a threat as it ever has, not because something new has happened biologically but simply because we have made little or no progress in dealing with it in fifty years. During those same years, the amount of natural rubber used globally has continued to increase, with demand expected to continue increasing over the coming decades. The rubber exported annually from tropical Asian countries weighs more than 8.5 million metric tons and is worth more than $20 billion per year, roughly twice the gross domestic product of the entire country of Laos.[16] By this measure, the rubber plantations are a country unto themselves, an empire of threatened trees.

12

Why We Need Wild Nature

The destruction of genetic resources is caused primarily by
the very success of modern plant breeding programs.
—Jack Harlan, "Genetics of Disaster"

Henry Ford imagined that farming trees was like running an
assembly line or a town or a state. He misunderstood the
strength of his rules relative to nature's laws. He misunderstood
nature's inevitabilities, the way pathogens find hosts much as water
finds a way downhill. In planting dense trees in the Amazon, he
took away the rubber tree's means of escape. It escaped by being
rare and hard to discover; it escaped by maintaining a distance
between itself and other trees. Today, in a world of commerce and
traffic, a world Ford himself helped make more connected, distances
have shrunk. Asia is no longer so very far away from the Amazon.

Leaf blight will arrive in Asia at some point. How will it come?
The spores of this fungus are thin and so don't do well on extended
travel, such as on boats, but they'd do fine on a plane. Knowing
what could happen, the Malaysian government prohibits planes
from traveling directly to Malaysia from places such as Brazil, where
leaf blight exists. But the bad news is that the spores may no longer
need to travel from the Amazon; they may have found a way around
the travel ban. They are rumored to have already arrived in Thailand. Many flights go from Thailand to Malaysia. Worse yet, as a
2012 study notes, "The pathogen can be easily isolated from infected
rubber trees...and transported undetected across borders,"[1] which

is to say that the intentional destruction of the majority of the world's rubber supply would be easy, just as the intentional destruction of cacao would be. It would be easy because the trees are planted densely; because most of the plantations are relatively close together; because the trees are genetically very similar to each other. It would be easy because the trees in Malaysia have not been selected for resistance; they have been selected for productivity. Planters chose trees with lots of latex, favoring short-term benefit over long-term security. It would be easy because we make the same mistakes again and again. It is for good reason, then, that the United Nations classifies South American leaf blight as a biological weapon.

Scholars express concern about whether terrorists might have the technology necessary to spread leaf blight to Asia. Do they have

Figure 9. Global rubber production per unit area by country (2005–2014). Dark countries are those in which leaf blight is now present. Light countries are those in which leaf blight is absent—for now. *Data source: FAOSTAT. Figure by Lauren Nichols, Rob Dunn Lab.*

the specialized knowledge necessary to transport and propagate fungal spores, the specialized knowledge necessary to destroy the world's supply of rubber? Of course they do, because all it would really take is a pocket full of infected leaves.[2] It could happen next year. It could happen in a hundred years. If it happens in a hundred years, we have time to plan and grow new trees, but given history's example, we are unlikely to do so. Instead the most likely scenario is collapse followed by global panic and then, maybe, a solution.

. . .

If we do nothing else to save rubber, when leaf blight arrives in Asia used rubber will be recycled, but recycling has its limits. During World War II, wartime regulations resulted in 65 percent of rubber in the United States being recycled, but it took less than a year before recycling was no longer sufficient. Then what? It is easy to dream up dystopian scenarios. Planes will continue to land for a while, until their tires wear. Airlines will have to make a choice between using unsafe old rubber and grounding planes. New cars will be built, but eventually the natural rubber necessary for sidewalls will be too expensive. Roads will eventually go quiet. Until what? Until, most likely, war—unless someone produces a miracle solution. The United States is the largest importer of rubber from Asia; China is the second largest. Were we to lose Asian rubber, or were it to begin to disappear, perhaps China would attempt to monopolize what remained. Perhaps the United States would intervene.[3]

. . .

One hope for preventing the worst-case scenarios is to have someone search for new varieties of rubber—or new genes from rubber— that confer resistance and then find a way to create from those raw materials varieties that both produce lots of latex and don't die. One hope is, in short, to do what Schultes and Imle were trying to do. In this regard, the story of rubber is yet another round in the cycle of crop failures that goes back to the very beginning of agriculture. But unlike the case of wheat, cacao, and potatoes, no "traditional" rubber farms exist where we might find seeds to conserve or knowledge on which to build. Hope rests entirely in the wilderness, the same wilderness from which the leaf blight sprung in the first place. As a society we might ask, "Why should we conserve wild nature?" As answers, we could talk about the aesthetic value of nature or its intrinsic value (we should save it because it exists, because it has a right to exist). We might even talk about the ecological functions of nature, for example, as a means to store the excess carbon we continue to release into the environment. But the most important answer is a simple one. Nature is not in balance, nor is it benevolent. It both threatens us and can save us. But whereas the dangers of nature, its pests and pathogens, will be able to threaten us no matter how much wild land we destroy, the benefits of wild nature are only available if we save the wild lands in which they live. The trouble is we don't know which wild lands hold the species we most need, either in general or in the specific case of rubber. We don't know because we haven't studied them, haven't scratched the surface in terms of understanding wild life.

Many, perhaps most, of the forests Richard Schultes visited are gone. Some of the unique populations of trees may be gone as well. A 2015 study estimates that half of all Amazonian tree species, including species of *Hevea,* are endangered as a result of the messy mix of habitat loss and climate change.[4] And unique populations and subspecies of trees, including those of *Hevea,* are necessarily rarer than the larger species they comprise. No one yet knows for sure how many unique populations of *Hevea* exist or even for sure how many species there are or exactly where they can be found. The forests of the Brazilian state Rondônia are (or perhaps were) a hot spot for unique rubber trees. But much of Rondônia has been cleared to make way for "progress," which often takes the form of soybeans. Rondônia is but one example in a world in which many once immense tropical forests are now just fragments. How much has been lost? No one knows.

It is possible that many of the populations of *Hevea* that Schultes found are still out there, for now, waiting to be collected anew and studied. Some may be hanging on in patches of a few trees, others flourishing. No one has gone through Schultes's notes and papers to retrace his steps in an effort to find the most promising species or unique populations. If you are looking for an adventure, this would be an exciting and rewarding trip. Nor has anyone tried to see if any knowledge from Imle's breeding program remains. Were his grafted trees really as resistant and productive as he contended? The few attempts to go back through the Amazon in search of seeds have been, relative to the scale of the potential problem, modest.

In 1981 a project was organized by the International Rubber Research and Development Board (IRRDB) to collect seeds in the Brazilian states of Acre, Mato Grosso, and Rondônia. Around sixty-five thousand seeds were collected and planted in Brazil, Malaysia, and Ivory Coast (fewer seeds than were collected by Schultes—or, before him, by Wickham—yet this was a lot of work). These seeds planted in Asia and Africa seem to have amounted to little. The seeds planted in Brazil all died or were lost. A handful of resistant strains are still being actively studied, some of which may be grown from seeds collected by Schultes (though this is hard to discern), but funding is modest and episodic. So far no varieties are both resistant and as productive as the varieties being grown in Asia.[5] As those who are still working to breed rubber resistant to leaf blight are quick to note, until there is a crisis, no one takes such efforts very seriously.

Among the only other attempts to do something different to save the rubber industry is work being done by a man named Eric Mathur out in the deserts of Arizona. Mathur has built greenhouse after greenhouse and filled them with unusual organisms, including bluebottle flies. The flies move back and forth inside the buildings; they are part of his dream. Mathur has an entirely different solution to the rubber problem, to which we will return shortly, but before we do it is useful to consider the general ways in which nature contributes, over the long term, to our crops.

. . .

Superficially, agriculture, including the farming of rubber trees, is often in conflict with wild nature. Ford cut down forests to make Fordlandia and Belterra. Similarly, the soybean fields of Brazil compete for space with the rain forests they replace. Any land being farmed is land that is not being conserved. Some kinds of agriculture are less in conflict with nature than are others, but none is without consequences, not now that humans number in the billions. Nor do the effects of agriculture on wild nature end at the edges of farms. The agriculture of the Green Revolution often leads to pollution— from pesticides, fertilizers, and herbicides—conferring negative effects that have a broad reach. Yet the sustainability of agriculture, our ability to keep farming beyond the next fifty, hundred, or even thousand years, depends upon the very forests threatened by farms. We need enough land in farms so that we can continue to eat (or, in the case of rubber, drive) in the short term, but not so much that it threatens our ability to eat in the long term. Although we don't treat it as such, nature is still the ultimate reservoir of our agricultural ingenuity.

The value of wild nature to agriculture is that it acts as a buffer for our ignorance. Sometimes we do a poor job of managing our crops, or evolution surprises us, and when this happens we turn to wild nature for solutions. In the context of rubber, because no traditional varieties of domesticated rubber trees exist, our dependence on wild nature is direct, mediated only by scientific knowledge and the traditional knowledge of indigenous Amazonians about rubber plants. We will need wild nature regardless of the severity of the problem. In the case of other crops, the dependence on wild nature emerges in the context of new extremes, whether of pests, pathogens, or climates. Sometimes the traditional varieties of crops are simply not varied enough to solve all our problems.

Sometimes agriculture benefits from wild nature through chance and sex. Liaisons between crops and their wild relatives move novel genes into agriculture. Nearly all such unions yield seeds and plants that are less useful than were their domesticated parents, but every so often a seed produced through such an exchange will have a new, useful combination of traits. Farmers have long benefited from the value of wild relatives of crops to their seeds. Traditional farmers in the highlands of Mexico, for example, are said to plant their corn crops near teosinte in order to infuse the corn with new traits, particularly when it faces new weather patterns or pests.[6] We can detect the history of such flow in the genes of our crops. We know from the genes of domesticated grapes that as farmers moved grapes west through Europe, their grapes bred with wild grapes. Rather than tossing the resulting fruits out, farmers planted them. We have Pinot Noir and Traminer as a result.[7]

Pollen can move from teosinte to corn in the wind. Grapes require insects to carry their pollen from flower to flower. Squash plants depend on a specific bee, the squash bee, for pollination.[8] Potatoes rely on bumblebees for pollination, and cacao relies on flies. But for many species scientists must do the work of the birds and the bees.

With time, breeders have become better and better at using wild relatives of crops in breeding programs. Such efforts offer huge benefits. Consider rice. Green Revolution rice varieties did well for around a decade, until new pathogens, particularly grassy stunt virus, started to show up. No domesticated rice varieties were resistant to grassy stunt virus, even though many hundreds, and by some count many thousands, of varieties were tried. The solution was found in the wild. The wild rice *Oryza nivara* is resistant to grassy stunt virus, as were the plants that resulted when it was bred with highly productive domesticated varieties of rice. The genes of *Oryza nivara* are now present in much of the rice you consume.[9] In the 1980s another pathogen affecting rice emerged, this time a bacterial pathogen. In that case a different wild rice, *Oryza longistaminata,* saved the day; one of its genes can now be found in some of the rice in your grocery store.[10]

Such examples exist for most crops. When phylloxera (*Daktulosphaira vitifoliae*), a root-feeding insect, invaded Europe and destroyed nearly all the vineyards from England to Albania, the solution came in the form of wild American grapevines. Nearly all wine in the world is now produced from grapes grown on the stems of domesticated grapevines grafted onto the roots of wild grapes. When corn was threatened by corn blight, resistance genes from one of corn's wild Mexican relatives, *Tripsacum dactyloides,* were bred into domesticated corn. In the last few decades, the genes of crop wild relatives have been crossed into bananas, barley, beans, cassava, chickpeas, corn, lettuce, oats, potatoes, wheat, sunflowers, tomatoes, and at least seventeen other crop species.

Wild nature saved corn. It saved rice. It saved wine—wine, for goodness' sake. It may well still save rubber. The annual value of the wild relatives of crops to agriculture in the United States is estimated to be $350 million. The global value may be in excess of $100 billion annually and is certainly not less than $10 billion.[11] More than any fuzzy but inconsequential panda or rhinoceros, the often scraggily wild cousins of our crops, in rain forests and grasslands, offer reason enough to save all the wild places we can. The value of these species will only increase.

As the world becomes warmer and less predictable as a result of climate change, agriculture will need to adapt. Some regions will become increasingly dry or experience an increasing number of unpredictable patterns. Irrigation can buffer these new extremes, but only if water for irrigation is available. For many small-scale farmers, crops are primarily rain-fed,[12] meaning that irrigation would require a new agricultural model, not just more water. Traditional varieties of crops from extreme climates may often have the genes we need to breed into farmed crops so that we can continue to grow them. And often the wild relatives of our crops grow in hotter and drier places than do any of their domesticated relatives, and so they are more likely to have genes that help as we continue to try to farm in marginal climates. For example, one wild variety of rice (*Oryza officinalis*) flowers earlier and later in the day than do most domesticated rice varieties. In doing so, this wild rice avoids flowering during the hottest part of the day. Recently the genes of this wild rice have been bred into a domesticated rice variety, allowing it to perform better in very hot conditions.[13] Another wild rice appears to be useful in breeding rice able to withstand the high levels of salt expected in many soils in the future.

The key to relying on the wild relatives of crops in the future is that these wild relatives need to still exist. One in five plant species on earth is threatened with extinction,[14] even before one accounts for the risks posed by climate change. The risk posed to the wild relatives of crops is likely to be even greater, inasmuch as they are concentrated in regions that have long been altered by humans (almost by definition) and so face the very direct threats posed by the same human populations they help to sustain. In a world in which we face many challenges, it may seem like someone else's job to support funding for conservation in Afghanistan, Turkey, and Iraq, yet we need to do just that if we want to continue to have bread to butter in the future—and beer brewed from barley with which to chase the bread.

Focusing on conserving crop wild relatives assumes that, over the course of the long history of agriculture, we have already found the most useful crops out there and that our task is just to keep them going. But we have not found them. It's possible that the species of most use to humans, even in terms of crops, are not yet being farmed. Therefore we need to conserve not only the wild relatives of crops but also species that might become crops in the future. Let's consider, once more, rubber.

. . .

By Schultes's estimation, no fewer than thirty-seven thousand species of plants produce latex of one form or another. More than seven thousand of those plant species seem to meet the most basic biological requirements necessary to be used to make rubber. We must conserve these species, too, though doing so vastly expands our list of species in potential need of conservation. By extension, it means that we should conserve species with potential uses other than their current ones. The species with tubers we have not yet domesticated. The species with fruits we have not yet tasted.

Several research groups have started to experiment with some of these seven thousand species. One of the species that has received the most attention is guayule (*Parthenium argentatum*). Guayule is a desert plant found in the southwestern United States and Mexico, a scrubby shrub with a woody stem. It is now being grown on a relatively large scale and sold commercially by two companies, PanAridus and Yulex. Both companies hope they can use guayule to replace rubber trees—or, if not replace them, compete with them.[15] The chief scientific officer of Yulex, Eric Mathur, claims to have bred more than twelve hundred hybrid varieties of guayule.

Mathur, whom Cade Metz described in a *Wired* magazine article as tall, dark, and bald,[16] has big ambitions. He crosses guayule varieties of interest in greenhouses. The choice of varieties is technological: he relies on new approaches to decoding the genetic differences among plants, then he chooses varieties with unique genes that when crossed are most likely to produce useful offspring. He has tools of which Norman Borlaug could only have dreamed, and the tools are getting better with time. For example, the USDA is decoding, nucleotide by nucleotide, the entire genetic code of guayule. But guayule flowers are picky, and so the critical element of the cross— the actual movement of pollen from the male parts of one flower to the female parts of another—is done using bluebottle flies. These glimmering blue-green flies carry pollen from one plant to another. They carry pollen among types of guayule long separated in the wild by geography, populations that were as varied, Mathur thinks, as any that existed. Or at least as varied as those that exist today. Guayule was harvested heavily in the 1900s, and during that time many populations of it were lost. Mathur is working to try to find some of these lost populations from old seeds in collections. Others, though, may simply be gone.

Some of the hybrids Mathur has bred produce a lot of rubber. Not yet as much as Asian rubber plantations, but enough to be potentially commercially viable. In addition, because of their small size, the plants and their latex can be harvested with tractors. Guayule is the great new hope. It is the hope of Mathur and Yulex; it is the hope of the other company working on it, PanAridus; and it is even a hope of the Cooper Tire & Rubber Company, which sees potential in the plant. So far, while you can find Yulex rubber in some products, it is not yet common. Patagonia, for instance, is now producing wet suits made of a mix of neoprene (artificial rubber) and guayule rubber. Bridgestone is producing some tires using guayule. Mathur and the others working on guayule are not the first to try using something other than rubber trees.[17] Others have failed where he hopes to succeed. Maybe guayule is a fool's errand, a dreamer's folly. But unless we radically change our approach to saving our crops, it is just the sort of errand on which we will continue to depend.

. . .

In theory, one could simply gather the seeds of all the wild relatives of crops and all the species that might someday have some use and save them in seed banks. You could store them until someone like Eric Mathur comes calling. In this way, saving the life around us from ourselves is an old sentiment. It is part of why, in addition to collecting 187,000 varieties of crops, Nikolai Vavilov also collected forty thousand different types of wild plants, most of them wild relatives of crops.

Yet although storing wild species in root and seed collections is one way to prevent the absolute loss of the wild relatives of crops, it can't solve all our problems. The seeds of some wild species simply do not store well even when frozen, or are hard to work with when they need to be replanted after storage. In addition, one must collect not only one of each species of a crop's wild relative, but also a representation of the genetic variation in each species. But the biggest problem is that once brought into collections, species stop evolving, which means they will never evolve to be able to deal with new pests. They are frozen in their moment of harvest. If they are to be fully useful, we must also conserve these plants in the wild in addition to whatever we save as seeds. It is in the wild alone that these relatives continue to meet the challenges tossed at them by nature.

Wild relatives are still evolving—unlike our domesticated crops— thanks to sex and pestilence.[18] Natural selection favors individuals in each generation who have the genes that will lead them to produce successful children. It disfavors, with the blunt ax of extinction, those who do not. Among the most powerful agents of natural selection are parasites, pathogens, and long-term changes in climate. Sex helps make sure that each generation is diverse, giving offspring genes from both parents, mixed in new combinations each time (with mutations scattered here and there for extra flavor). Sex engenders diversity, and natural selection winnows down the resulting forms. If we are to continue to generate new crops, crops that are resistant to climate change, pests, and pathogens, we must save the places where changes in climate, new pathogens, and new pests engender new varieties of crop relatives. We must save not only our crops' wild relatives and the species they depend on but also the species that attack them—the pests, pathogens, and parasites that drive their evolution.[19]

Ultimately, the more intensive agriculture becomes, the more important wild relatives of crops will become. The more extreme new climates become, the more important wild relatives of crops will become. The more new pathogens emerge, the more important wild relatives of crops will become. To make ourselves safe, we need to save wild nature, and we need to do so on a large scale. Conservation biologists have developed sophisticated models of how much and where we need to save in order to save ourselves. But E. O. Wilson, who has spent a life in the trenches trying to conserve biological diversity, offers a simpler model. We need, he argues, to save half the land. In some places that half is more important than others, yet the need to save half is a recognition of a simple truth, that saving wild places for the uses they will one day provide is worth at least as much as whatever we can do or grow for ourselves today, right now. What we farm now—it is for us, today. What we save—it is for our children and grandchildren forever, the biological gold they will prospect. Save a forest today and it will help to pollinate coffee. Save a forest for tomorrow and it will help to keep coffee from going extinct.

Conservation biologists talk a lot about reasons for conserving wild species. Some are aesthetic (they are lovely). Some are intrinsic (they have a right to exist). But even if you're unconvinced by those arguments, it is hard to deny that we need to save wild places and wild species if we plan to save ourselves. We need to do this in order to continue to eat, drive on rubber tires, and exist in the civilizations we've become comfortable existing in. We should want to save wild places for their own sake, but if we don't care about the innate beauty or value of the wild—the wild rubber, the cacao bee, the bluebottle fly, and guayule—then we should at least conserve them to save ourselves.[20]

13

The Red Queen and the Long Game

> It can be hard to tell a crank from an unfamiliar gear.
> —Leigh Van Valen, as quoted by Douglas Martin in Van
> Valen's *New York Times* obituary

Sometimes I consider writing a novel about the dysfunctions of a biology department. The more true to life I would make it, though, the harder it would be for the reader to believe. Some of the dramas of academic departments relate to inability of "great scientists" to deal with ordinary problems or at least to do so in ways that society judges to be appropriate. A professor might, for example, pee in a jar in his office in the name of efficiency, even though the bathroom is just a few hundred feet away.[1] Others relate to the problems of specialization. Biologists have become so specialized that individuals within a department can come to think that what others in the same department are studying is not interesting or, worse, not even science. Then there are the human dilemmas made more unusual for their setting—a love drama set in the room where collections of preserved brains are kept in glass jars of formalin, for example. But what makes biology departments most interesting is that amid their passive-aggressive theater of egos and failings, real truths about the biological world and how it works are being revealed, even by those who most conspicuously walk to the beat of their own tambourines.

To envision him properly, you need to know that Leigh Van Valen had no proclivity toward ironing.[2] His perpetually wrinkled shirts hung differently on the left and right sides of his body because his right pocket was always filled with three-by-five index cards. He took notes on the cards, tiny notes. If you were saying something interesting, he took notes about that; if you weren't, he took notes on the more interesting conversation going on inside his head. He has been described as having had a beard slightly longer than God's but slightly shorter than Charles Darwin's.[3] His office was a labyrinth of shelves and tens of thousands of books balanced in precarious towers. It was an edifice of words and ideas amid which he composed songs about the sex lives of dinosaurs and how the subject of his adoration compared to a lemur (unfavorably, as it turns out).[4] He pushed the extremes of ordinary behavior even in an academic department. He also laid bare the fundamental rules by which life works.

During the 1970s, Leigh Van Valen was a professor in the department of ecology and evolution at the University of Chicago.[5] He was an evolutionary biologist who studied many facets of evolution.[6] He didn't like to study things others spent time on or anything likely to get funded. "Conformity stultified," he wrote about science, though surely it was also a broader assertion.[7] Beginning in the 1960s, Van Valen used math and theory to make predictions about the general rules by which the real world works. He was particularly interested in the struggle of hosts, be they humans or agricultural plants, to escape their pests, pathogens, and predators.

No species, Van Valen noted, is immune from parasites, predators, and the need to get food. As a result, no species in nature ever perfects anything. Each species is constantly running toward food and away from its demons, and so any lurch forward, to escape or catch, is temporary—the act of raising a foot that will, as soon a predator or parasite adapts, fall in the same place it started. Van Valen called this idea the Red Queen hypothesis. The name of the hypothesis is borrowed from the imaginings of Lewis Carroll in *Through the Looking-Glass*. Alice asks the Red Queen why she runs and runs but seems to stay in same place. The Red Queen responds that to advance, she must run twice as fast. The Red Queen is a character that speaks to our human condition. The Red Queen, Van Valen noticed, was also a metaphor for a common scenario in life, especially the life of agriculture.

His hypothesis was, Van Valen thought, new and important. His peers were not so sure. When his paper setting out the hypothesis was rejected repeatedly, Van Valen started his own scientific journal and named himself as its editor. In that journal, the article was immediately accepted. It was the first article in volume 1 of *Evolutionary Theory*, published in 1973.[8] The quantitative aspects of the paper would prove elegant in hindsight. But it was not the math that gave the paper its lasting effect on the field; it was the metaphor of the Red Queen. In referencing her Van Valen allowed us to think about the living world in light of an example we could envision. A great analogy is itself a kind of discovery.

· · ·

The whole history of agriculture can be recast as a series of attempts to get a little farther ahead only to realize that we are staying in place. Initially the Red Queen race on farms was similar to that in grasslands and forests. On traditional subsistence farms, farmers— escape artists, in a way—have a variety of ways of staying just ahead of pests and pathogens. They grow multiple varieties of individual crop species and multiple species of crops at the same time so that the odds of at least one of them being resistant to a new pest or pathogen are higher. Such polycultures also have the additional advantage that they buffer farms against other uncertainties, such as those due to year-to-year differences in climate. In addition, in light of such diversity, no pest or pathogen has enough food and is therefore unlikely to cause an outbreak. In such fields, many crop varieties are running from many pathogens and pests at the same time. If the varieties of a particular crop, such as wheat, in the field are not enough, farmers trade and borrow in order to keep running, which is possible to the extent that the Red Queen race is happening on each farm in each village independently. The resistant varieties, once found, are then grown in large quantities, which leads the pathogen to become scarce until it evolves some new trick and once more becomes common. Traditional farms escape pests and pathogens by being diverse at any moment across space and time.[9] Theirs is a complex dance similar in many ways to the one going on in natural grasslands and forests, a many-partnered tango of death and sex.

For most crops, however, the biggest jump in the Red Queen race, as we've seen many times, occurred when they were moved and, in being moved, experienced escape from their pests and pathogens. But eventually, as we've also seen, so many pests and pathogens catch up that the benefits of escape are gone. Some crops still grow in this luxurious reprieve, but their numbers diminish each year.

After the crops were moved and the pests and pathogens started to catch up—which appears to have happened first in temperate regions, where rates of movement within certain climatic zones (e.g., from Europe to North America) are greatest—the next step was to breed resistant varieties of crops. Even before the Green Revolution, these resistant varieties washed over North America, Europe, and Asia with great speed. As a result, world food supplies were 20 percent higher per person in 2003 than they were in 1960, even though the global population more than doubled[10] and even though the Green Revolution only really affected farms in a handful of countries. With these new varieties, agriculture got ahead, temporarily. The greatest yields also began to become concentrated in affluent

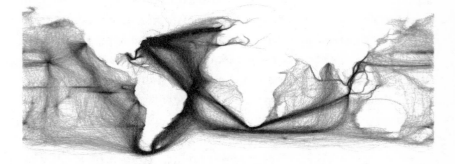

Figure 10. Global shipping routes. Note the extent to which the Canary Islands remain a hub for global shipping. While not all routes are equally trafficked, pests and pathogens have many opportunities to get from one place to another, and the rate and extent of global shipping is increasing. *Image source: Benjamin Schmidt.*

countries—countries in temperate regions, such as the United States. However, these yields came with a catch. Their use sped up the race.

．　．　．

Often the resistance of crop varieties to pests and pathogens comes from a single gene or trait that confers defensive capabilities. But if all the plants in a field (or, worse, on a continent) have the same defense mechanism, any pest or pathogen that can overcome the defense can devour them all, which is what happened next. Southern corn leaf blight showed up first in Iowa and Illinois in 1968, where it caused "ear rot" on ears of corn. At the time the blight was not yet named; it is now known as *Bipolaris maydis*. In 1969 it was more common, but still not considered a real problem. Then in 1970, southern corn leaf blight swept from one side of the country to the other, a gash in the cornfields that caused $1 billion in damage—in 1970s dollars. Fifteen percent of the corn in the United States was destroyed; in retrospect it is unclear why it wasn't more.

Southern corn leaf blight moved well because 1968, 1969, and 1970 were moist. It also moved well because of the lack of diversity in corn varieties. But most of all, it moved well because even though several varieties of corn were grown in some regions of the country, all the corn in America had the same weakness, thanks to the Green Revolution. A variety of corn had been found in which the male parts of the corn were sterile. Corn can "self," which is fancy scientist speak for the ability of some plants to have sex with themselves. Selfing is a problem for breeders trying to produce hybrids. Historically the way around selfing was the removal of the male parts of the corn flower—detasseling. But this must be done by hand. The sterile male corn was great for plant breeding because it gave breeders control over which varieties bred with which other varieties. It was, in turn, great for farmers because it allowed them to achieve better yields with less labor. The sterile male corn meant farmers did not have to do the hard work of detasseling any more. And there shouldn't have been a problem, because sterile male corn could be used to produce multiple varieties that were identical in one and only one feature: the genes in the cytoplasm (the tiny sea within each cell) that made the males sterile. The problem was that if a pathogen ever figured out how to manipulate those genes or what they made, nearly every ear of corn in the country would be in big trouble. That is just what southern corn leaf blight did. In less than a decade it found the Achilles' heel of the corn industry; the silver bullet had become the Achilles' heel. What's more, having devastated the corn industry in the United States, southern leaf blight spread to the other regions in which Green Revolution corns were growing. It may have been spread (knowingly, some argue) by seeds shipped out of the United States during the years of the blight. It was soon in Japan, the Philippines, parts of Africa, and Latin America. Everywhere southern corn leaf blight arrived, the corn failed.

To pests and pathogens, the Green Revolution presented a new kind of bounty. Once a pest could escape whatever held it back, it could eat for miles, unaffected by the subtle differences among crop varieties, unaffected by the presence of its enemies. If there is an Eden, it is what these pests, once they escaped the defenses of the Green Revolution crops, encountered. The consequences of such escape are predictable[11] and yet still hard to deal with. Once a pest showed up that could eat a crop variety, one choice was to replace that variety with another. This is what happened with corn. It is also what happened with sugarcane. For sugarcane the only recourse has been to introduce a new variety every ten to twelve years, one the pests can't yet deal with. Varieties lasted just three to four years, on average, before the pests caught up and new varieties had to be planted. (Ironically, the race is fastest in crops in which breeding is most active. In these crops, new varieties tend to be higher yielding, leading farmers to switch crops both when new pests arrive or evolve and when new seed varieties yield more than the old.)

The next hope for slowing down the race between new crop varieties on the one hand and pests and pathogens on the other was investment in new chemicals, pesticides, herbicides, and fungicides. These compounds allowed agriculture to jump ahead. Again, investment was concentrated in relatively few and relatively affluent countries. But of course even in the face of affluence, nature quickly caught up again in the form of resistant insect species and pathogens.

Records of insects resistant to pesticides go back to the late 1800s, when pesticides were still made from mixtures of heavy and toxic metals. But initially these records were rare and of relatively modest consequence. The first record of pesticide resistance was reported in 1914 to sulphur-lime, a deadly kin to copper sulfate. Then, in 1928, the codling moth (*Cydia pomonella*), was found to be resistant to lead arsenate. Over the following thirty-eight years, other insects also evolved resistance, but so few that modest innovations in pesticides seemed likely to be sufficient to allow agriculture to keep up.[12]

Unfortunately, as Norman Borlaug's new crops spread and crops dependent on pesticides were sprayed in greater quantities, the rate of evolution of resistance to pesticides increased. Consider the case of DDT. DDT was first used in 1945. In less than seven years, resistance was detected. By 1954, several species of insects evolved resistance to DDT every year. By 1980, one or more populations of each of one hundred and six crop pest species had figured out the secret handshake. The rate sped up because larger areas of crops were being planted and more of the pesticide was being used. The first arsenic-based pesticides lasted more than sixty years before resistance was observed. Dieldrin lasted three years, Endrin two years. Carbaryl four years. Azinphos-methyl five years. Then things got worse. None of the pesticides developed between 1973 and 1979 went more than two years before resistance emerged. This was to be the new norm. We were only just barely keeping up, certainly not running ahead. The cycle went on from crisis to crisis. Instead of relying upon diversity within a field to delay resistance or virulence from evolving in a pest or pathogen, the new model was to wait until resistance evolved, then look for a new pesticide. Or if the resistance was in the crop itself, wait until a pest figured a way around the resistance genes of the crop, then plant anew. The new model was brazen. It was a sword of Damocles over our dinner plates. The repeated emergence of pests and pathogens resistant to crop varieties, herbicides, and pesticides wed farmers even more tightly to the companies that sold seeds, herbicides, pesticides, and fertilizers.

The more the race depended upon breeding resistant varieties of crops or formulating new pesticides, the less individual farmers were part of the race. As a result, an increasingly global swath of farmers depended on an increasingly small number of companies, universities, and institutes to keep pathogens and pests at bay. The burden of helping crops escape pests and pathogens shifted from tens or even hundreds of thousands of farmers to a far smaller number of scientists. And those scientists would have to bet on the blind hot light of their ingenuity, often forgetting that such ingenuity frequently comes from looking to traditional crops or wild species for clues. And there was more. Whereas the Green Revolution started with university research and research done by institutes like that for which Borlaug worked in Mexico, its success, along with changes in global governance, began a process in which federal governments and universities in affluent countries spent less on agricultural research in their own countries,[13] international development agencies spent less money on agricultural research in developing countries, and, day by day, the onus of keeping crops growing shifted from public institutions to private ones. This was to have enormous consequences, though perhaps not the sort that anyone anticipated.

In theory, the Green Revolution could have been one in which we relied on an understanding of ecology and evolution to increase crop yields while at the same time keeping crop fields diverse. Research has shown that modern intensive rice fields that contain diverse varieties are less burdened by disease and more productive than those planted with fewer varieties.[14] Varieties susceptible to a particular pathogen suffer less from that pathogen when surrounded by varieties that are resistant. The trade-off is that diverse fields are harder to plant because seeds are often non-uniform in size and so less amenable to standard machinery built to sow grains. They are also more challenging to harvest because the plants grow to different heights or mature at different times. These challenges might have been overcome. But they weren't, especially in developed countries that increasingly relied on mechanization in agriculture after the Second

World War. Instead the Green Revolution took us down another path, one in which we must rely on technology and industry to help us run ever faster in order to stay in place.

. . .

Elsewhere, another approach to agriculture was starting to emerge, something that provided a way of permanently escaping pests and pathogens. It started with a moth. The flour moth (*Ephestia kuehniella*), which you are very likely to have in your own kitchen, is, like most animals, always host to beneficial microorganisms.[15] It is also, like most animals, sometimes host to pathogens. Among the latter was an especially destructive species of *Bacillus*. Named for the region in Germany in which it was discovered, Thüringen, it was to become *Bacillus thuringiensis*, a.k.a. Bt. It was, it would turn out, a bacterial lineage that in Japan also assailed silkworms—the caterpillars used to produce silk.[16]

In your kitchen, *Bacillus thuringiensis* helps keep the moths living in flour in check. But in the early 1900s in Japan, *Bacillus thuringiensis* was a major problem for the silk industry. Bt produces crystallized proteins that, once consumed by the silkworm, are broken down by enzymes in its alkaline gut. These activated proteins, freed from the crystals, bind specifically to proteins attached to the gut like keys in locks, creating tiny holes that allow other bacteria, in addition to the Bt itself, to invade. Gruesome. If you are trying to raise silkworms it is a major problem, a brutal one at that. But it also quickly seemed to be an opportunity inasmuch as if the proteins from this silkworm bacteria could be produced in quantity, they could be used as a pesticide to kill insect larvae. The bacteria could be sprayed onto crops, where the insect larvae would consume them. The bacteria, just as they did in the silkworm, would then destroy the larvae that account for much of the damage to crops globally.

In the first trials, which began way back in the 1930s,[17] the bacteria killed European corn borers. By 1938, a product based on the bacteria was on the market in France. By the 1970s, after several fits and starts in further development of the bacteria for commercial use, the bacteria were being raised in huge vats. In the vats, the bacteria produced the crystallized Bt proteins (or Bt toxins, as they are more often called). The bacteria themselves were then used as a pesticide.

Because the pesticide was produced naturally, by bacteria, it could be used on organic farms. Even better, as scientists started to search for these bacteria in nature, they found them everywhere—many varieties producing hundreds of different (but related) kinds of protein crystals that appeared to affect various groups of insects. Most were effective against caterpillars, but *Bacillus thuringiensis israelensis,* for example, was effective against mosquitoes. Another variety kills Colorado potato beetle larvae, a near miracle. Whereas the search for new pesticides through chemistry has a very low "hit rate"—roughly one in twenty thousand has any value—the hit rate for the Bt toxins was one in every one thousand. What's more, because this pesticide was naturally produced by the bacteria, bacteria at war with insects, it might be harder for insects to evolve resistance to it, particularly if different versions of the proteins were mixed and matched. It worked; organic agriculture experienced a boon. It was a boon based on nature, contingent on the fact that that nature was there in the first place.

Suddenly scientists and companies began to search the world for these bacteria. They found them everywhere, though they were, it seemed, most common in grain silos and in soils. Each type of soil seemed to have its own variety. The more basic kinds of organisms that are used in agriculture, the more value that can come from unknown and poorly studied wild species (if they are studied, if they are saved). No one had ever really thought about the conservation of bacteria. But here agriculture was benefiting from bacteria that had been saved, inadvertently. They were saved when forests and other lands were conserved in order to protect wildlife, be they spotted owls, black-footed ferrets, or redwood trees. Here was yet another reason to set aside half our land for conservation, to save the bacterial species we have not yet discovered, species that could change the way we live and farm.

· · ·

In these bacteria, agroindustrial companies would come to see an opportunity, a way to potentially escape pests and pathogens, perhaps forever. It would be a final, triumphant step in the history of agriculture, a moment in which the wilderness was defeated once and for all in favor of civilization (never mind that the species being used to further this goal came from that very same wilderness). It was to be a desperately necessary step, because short of something truly new we are stuck in a race against our pests and pathogens—a race in which we are hindered by a growing human population and changes in climate that will, on average, make farming harder. What if gene splicing, an approach developed by Stanley Cohen and Herbert Boyer in 1973, could be used to isolate the genes responsible for producing the Bt toxins and insert them into the genomes of plants? If this could be done, it would allow breeders to make any plant deadly to pests. One could choose the most productive plants in the world, then make them resistant to pests without having to spray them with pesticides.

Putting bacteria genes into crops would be a big step in the history of plant breeding, a wild monkey jump. The clever minds of science would have found a way to the future.[18] Throughout history, plants could only be bred if they could be made to reproduce with each other. Farmers could be creative in getting plants to mate, but they were subject to the biophysical realities of pollen and ovaries. Exceptions occurred naturally. For example, sweet potato genomes contain genes from a bacterium that somehow made the jump (and, in doing so, benefited the sweet potato).[19] But such examples are rarities, beyond the active control of any human. Yet in splicing, two organisms that could not reproduce could be made to swap genes, as the sweet potato and the bacterium did, and they could be made to do so relatively predictably. In the case of Bt, specific bacterial genes associated with toxicity to insects could, in theory, be inserted into a plant. If the plants could produce the toxins, caterpillars eating the plants would ingest them, and no pesticides, or at least fewer applications of pesticides, would be necessary. What was novel in terms of the history of agriculture was being able to direct and, ultimately, speed up and control the process of inserting genes from one species into the genome of another. Unrelated organisms could potentially be made to share genes, and as a result, all the species on earth—be they jellyfish, bacteria, or squirrel—could be of value to agriculture. Gene splicing would make wild nature more valuable to farms. What was novel from a business perspective was that if this worked, any given company could sell you the biological and chemical requirements for a farm in a single bag—in a single seed, no less.

Eventually the splicing of Bt toxin genes into plants worked, though only after many years of trial and error and at great expense. It was to be among the first attempts to genetically engineer a crop in the laboratory.[20] Scientists at the agricultural chemical company Monsanto did the work, building on discoveries made by university scientists. At the time Monsanto, located in Saint Louis, Missouri, had ridden to success on the back of the Green Revolution. In the act of splicing Bt toxin genes into crops, the fusion of chemistry and agriculture was coming to its most elaborate fruition. Monsanto could now sell its chemicals by selling seeds. Its chemicals were Bt toxins. Monsanto could offer the core of its agroindustrial portfolio embedded in a single grain. Just add tractors, irrigation, fertilizer, and herbicides.

Monsanto took seeds originally domesticated by traditional farmers and updated them using modern genetics. The seeds were still mostly the same traditional seeds, with a few edits that gave them the requisite superpowers. Other companies, nearly all of them originally chemical or pharmaceuticals companies, including Bayer, Syngenta, and Dow, soon followed suit. As they did, they purchased many of the major seed-supply companies in the world. Selling seeds and selling agricultural chemicals were no longer separate endeavors. It is hard to imagine they will ever be again. The good news, though, was that these seeds and their associated chemicals offered a way to get ahead—and, just maybe, stay ahead—of the pests and pathogens. What could possibly go wrong?

14

Fowler's Ark

In the near future, man will be able to synthesize forms com-
pletely unimaginable in nature.
—Nikolai Vavilov, *Origin and Geography of*
Cultivated Plants

If the tenuous bubble of civilization bursts, whether after nuclear
war or something more mundane, humans will find themselves
once again wandering the earth, looking for seeds.

Small seed banks are likely to be gone. With electrical grids shut
down, the seeds would succumb to rot or rats. (The one thing dystopian fiction writers and scientists agree on is that after the apocalypse, there will be a lot of rats.) For a while, the largest national seed
bank in the world, the one in Fort Collins, Colorado, would still have
seeds. But it, too, depends on electricity. The one seed bank likely to
be left is in Norway, on an island called Spitsbergen, in the Svalbard
archipelago. To get to the island from any mainland, one would have
to paddle a boat to the shore, then climb a mountain to a place
beyond the abandoned coal mines to the Svalbard Global Seed Vault.
A rectangular gray entrance protrudes from a hill, an entrance
designed to be recognizable as significant to future civilizations.
Inside, down a long tunnel in the earth—kept cold by the perma-
frost of the tundra, even without electricity—a few old botanists,
fellow survivors of the apocalypse, would still be tending to the seeds
from which crops and civilization might bloom anew. Perhaps the
scenario I have just described is far-fetched, yet it is just the sort of

scenario that the builders of the Svalbard Global Seed Vault had in mind. They built the vault to be used when all else is gone.

. . .

The dream of the "doomsday vault" has some of its earliest roots in 1971, when, at the age of twenty-two, Cary Fowler found himself in the hospital, naked but for an ugly gown. His doctor had found cancer. It was spreading fast. The doctor tried to remove the mass of dividing cells. But he was unable to get it all. The doctor told Fowler he had six months to live. He asked Fowler if he had life insurance. Fowler planned his days. He was dying, and what struck him most deeply was that in his short life he had never done anything great.[1]

Then something totally unanticipated happened. Fowler lived. His cancer disappeared without explanation. His immune system, the doctors said, had been good enough to distinguish his healthy cells from those of the cancer and to fight the cancer. He was still young, yet already he knew what it felt like to face death having accomplished little to change the world. He did not want to feel that feeling again; he would do something great. But how? And what?

Fowler tried his hand as a writer for *Southern Exposure Magazine*. He covered the story of the demise of southern farms. Across the southern United States, large farms, born of the Green Revolution, were buying out small farms. With this transition, a way of life was disappearing. On the basis of his work at the magazine, Fowler was invited to help do the research for a book called *Food First* by Frances Moore Lappé.[2] Fowler moved to Hastings-on-Hudson, in Westchester County, New York, where he lived in a rented house and worked side by side with Lappé — who was by then already well known as the bestselling author of another book, *Diet for a Small Planet* — and her coauthor, Joseph Collins.

Food First was an argument against the agricultural industry and the simplification of agriculture and in favor of the enrichment of life — through eating locally grown foods, maintaining a vegetarian diet, and emphasizing flavor over quantity. While helping with the book, Fowler started to read articles written by Jack Harlan. In reading Harlan's work, Fowler found his passion for seeds.

. . .

Jack Harlan, then a professor of genetics at the University of Illinois, worked his entire professional life to study and conserve traditional crop varieties (along with the wild relatives of those varieties). In Harlan's work, Fowler found as clear a statement as he had ever seen of his fears. Fowler did not fear death. He feared the lack of accomplishment. And he was realizing that, like Jack Harlan, he also feared the consequences of losing traditional varieties of crops.

The industrialization of modern agriculture, as Fowler had noted in an article in *Southern Exposure*, threatened many aspects of small-town life and small-farm agriculture. But what it threatened permanently, Fowler saw from Harlan's writings, were traditional varieties of crops and their seeds.[3] Much as the cancer cells in Fowler's body had competed with his own healthy cells, the industrialized crops of the Green Revolution seemed to be spreading around the world and competing with the native crops of the regions in which they arrived. And whereas the crops native to each region were diverse, the crops of the Green Revolution were simple—able, like the cancer cells, to grow fast at the expense of the wilder and more diverse tissues of life.

Harlan's writing referenced some of the ideas Leigh Van Valen would come to emphasize. This included a hypothesis I'll call Harlan's ratchet, which posits that while traditional crop breeding engendered an increasing diversity of crops, the new, commercial breeding did the opposite and that as a result the loss of diversity would be progressive.

The beginning of modern agriculture is simplicity: crops are bred with their close relatives until the resulting seeds (so-called pure lines) are homogeneous, nearly identical. Where new forms are created, they are made from existing pure forms. Said Harlan in 1972, "A large proportion of the genes of the old landraces was discarded" with each generation of breeding (a landrace is just a local variety of a crop).[4] This was true not only for staples but also for less intensively farmed crops, as long as they were being bred for large-scale production rather than flavor or suitability to local soils, climates, and other conditions. Between 1903 and 1983 the number of cabbage varieties one could readily find dropped from 544 to twenty-eight, carrots from 208 to twenty-one, cauliflower from 158 to nine. Genetic engineering of crops accelerates this narrowing. Genetic engineering almost always focuses on adding new traits to a single variety of, say, corn. It is expensive, and so it is hard for companies to justify working on any but the highest-yielding varieties.

Yet Fowler realized that Harlan's ratchet would lead to progressive decreases in the diversity of traditional crop varieties.[5] Here was the great irony of industrial agriculture. Green Revolution crops were produced by breeding many different traditional varieties. But the success and spread of Green Revolution crops was destroying those same varieties, draining the very reservoir on which its future would need to be sustained.

In 1978 Fowler took a job at the Rural Advancement Fund (RAF; associated with the National Sharecroppers Fund) in Pittsboro, North Carolina, about twenty miles west of Raleigh. RAF was established to aid black sharecroppers in the southern United States in the wake of the mechanization of cotton picking and the consequent loss of jobs. RAF staged protests and other civic actions against large farms and local and regional governments, and organized civic action on behalf of poor farmers. With time, RAF also took on other regional social justice issues, such as those associated with migrant labor. It was this latter issue that led Eleanor Roosevelt to begin to work with RAF. RAF did not work with seeds, nor did it concern itself with issues outside the United States. Then Fowler arrived.

At RAF, Fowler found others who shared his growing desire to make changes through activism. But whereas RAF focused on the United States, Fowler's vision was global. Rather than change his focus in response, Fowler decided to change RAF. He persuaded his colleagues that RAF should help poor farmers everywhere and that it should do so by saving their ability to use and benefit from their traditional seeds. Fowler turned RAF's mission into a mirror of his personal goals.

Relatively few varieties of seeds were domesticated in North America and western Europe. Instead agriculture in these regions depends on the crops created in the southern tropics and subtropics and taken northward by conquistadors and explorers (where they were added to the seeds brought west, thousands of years earlier, from the Fertile Crescent). Some historians have gone so far as to suggest that it is this movement of crops that precipitated the economic success of Europe and the United States in the last two centuries.[6] Even Vavilov's story could be seen as part of this narrative. His collection of seeds benefited Russia far more than it benefited any of the regions from which he gathered the seeds.

The very poorest farmers in the centers of crop diversity, whose seeds had fueled the agricultural success of modern Europe and North America, were Fowler's first priority. The easiest approach to saving the seeds produced by these people was saving seeds in one central place and then, having done so, helping develop regional seed banks as well as programs to help farmers in the field. This might have been easy at one point, but then, in 1970, the United States' Plant Variety Protection Act allowed sexually produced plants—plants produced through traditional breeding—to be patented; this would include the plants produced during the Green Revolution. In 1980, two years after Fowler arrived at RAF, a Supreme Court decision building upon this law also allowed new forms of bacteria and plants that had been produced asexually, including those produced through genetic modification in the lab, to be patented. These laws had upsides. They provided a framework wherein breeders could expect to benefit from the new varieties they produced, even if the seeds of those varieties only needed to be purchased once (as is the case with self-fertilizing plants, which produce offspring, a new year's seeds, nearly identical to themselves). But the laws meant that not only were the traditional farmers who produced the seed varieties the breeders were tinkering with in the first place not benefiting the most, in some cases they also had to pay for the very seeds their ancestors had created. As a result, farmers and countries in the hot and poor regions of the Earth became less rather than more likely to share their seeds.

To advocate on behalf of the farmers and countries whose seeds were being patented, Fowler decided to go to the annual meetings of the Food and Agriculture Organization (FAO), on behalf of RAF. Once there, he worked with the member countries of FAO to fight for a framework that would allow seeds to be shared and saved in a way that benefited the countries in which those seeds were first domesticated. Then in 1981, Fowler was diagnosed with cancer again. It was a testicular cancer unrelated to his earlier case, and it was growing fast; it was all through his body. The good news was that the doctor had treated similar cases twice before. The bad news was that in both cases the patient died. Fowler wondered again: had he done enough with his life? Whatever it was that he had accomplished was still not very concrete.

Fowler's treatment appeared successful. He was in remission from two different cancers. Once more reborn, he was emboldened. He went to every FAO meeting and continued to agitate. It was not long before Fowler's work was recognized. In 1985 he was awarded, along with Patrick Mooney (his friend and coconspirator), the Right Livelihood Award. It was a recognition of the work the two men had done to save the world heritage of seeds as well as their attempts to come up with a new and more just framework for agriculture. At the awards ceremony, Fowler and Mooney gave a speech. The speech was an opportunity not only to raise the concerns that worried them but also to offer a concrete plan, which they proceeded to do.

The plan had two key elements. A collective seed bank needed to be created under the flag of the United Nations in which seeds from around the world could be safely stored and available to everyone. The thought was that such a seed bank could be distributed among many regions, but where already successful seed banks might be leveraged to engender further successes. Sweden, for example, where Fowler had done his PhD, had a seed bank used by all of the Nordic countries. Under the UN banner, it could be used by any country. In addition, if countries made money from seeds first domesticated in other regions, they needed to provide funds to those countries to be used for studying their seed heritage and improving upon it. The wealthy countries, in short, needed to help support the heritage on which their successes were built. Fowler's immune system fought off his cancers using a set of rules—policies—about which cells to feed and which to favor and how. Similarly, Fowler believed that the key to maintaining a healthy agricultural system also lay in rules and policies.

Initially, with such proposals Fowler was a radical, disrupting the status quo, someone not invited to the party. He was a thorn in the side of agroindustrial companies, but he was also a thorn in the side of the United Nations, where he criticized the organization's inaction on the matter of the global heritage of seeds. Someone at the UN needed to do something, Fowler implored. His voice got harder to ignore in 1990, when he and Mooney wrote a book called *Shattering*,[7] which laid out both their critiques and their ideas in extended form. Finally, in 1993, having heard Fowler again and again say that someone needed to do something, the UN asked him to do it.[8] He was recruited by the FAO to help draft a plan for the global use and banking of seeds. In 2001, the UN adopted the proposal to which Fowler had contributed, which included both a plan for sharing seeds and a statement about the need for a global seed bank in which to store backups. It was to become a template for the International Treaty on Plant Genetic Resources for Food and Agriculture, also known more simply as the International Seed Treaty.

Under the treaty, farmers, breeders, and scientists all get access to the seeds produced via traditional means as long as they give an equitable share of their profits from the seeds they market back to the country that originally produced the seeds. It is not clear what an "equitable share" is, nor is it clear what happens if countries or companies fail to comply. Still, the treaty covers sixty-four of the most important food and forage crops. The choice of which sixty-four to include was, of course, a bit arbitrary (or at least political), but it's nonetheless a pretty long list, especially since ten or so of the crops on the list account for most of our food. And the treaty allows for the patenting of seeds. There had been no way around that—no way to come up with a treaty to which all parties would agree.

The treaty, after so many years of struggle, set the stage for the other part of Fowler's plan, a global seed bank. All Fowler's work had built, painfully slowly, up to this point, the point where he could really get started. But calling for a seed bank and making one are two very different things. Fowler began to try to figure out how to actually do it. In his life up to that point, he had accomplished many things—he had articles, books, and a treaty to show for it—but these things were all based on words. Now he wanted to make something concrete, a physical structure from which humanity might benefit. It was a different sort of task. He had ideas—he had always had ideas— but now he needed more than ideas. He also needed money.

. . .

Fowler was in a kind of race to build the seed bank while he was still healthy and before too much of the world's small-scale agriculture was gone. Meanwhile, agroindustrial companies were moving at their own, much faster, pace. The companies began to buy up private seed-supply companies. They could not own the seeds in public seed banks, nor could they patent and protect the seeds they had (or had engineered) as readily as they hoped, yet they could control as much of the diversity of seeds as possible. Monsanto, for instance, bought or otherwise took control of Holden's, Asgrow, De Ruiter, Carnia, Monsoy, Seminis, and several dozen other seed companies.

The new seeds of the agroindustrial companies, particularly the transgenic seeds produced by Monsanto, spread. The first crop into which Bt genes were inserted in order to kill pests was tobacco, but Bt tobacco was never commercialized. Bt potatoes were subsequently produced and commercialized, as were Bt corn and cotton. Large-scale farmers who could afford these crops liked them. Then new seeds were produced that were resistant to herbicides. These seeds could be sprayed with the herbicide glyphosate and survive. This did not reduce the cost of farming the plants, nor would it reduce herbicide use. It simply reduced the trouble required to control weeds in a field. Then seeds with both these traits—resistance to pests and resistance to herbicides—were created. Soon much of—and then nearly all—the soybean, cotton, and corn in the United States was genetically engineered to produce organic toxins and/or be resistant to herbicides. The United States was probably among the countries in the world that least needed extra calories from corn. Yet it was hurtling faster toward transgenic crops than

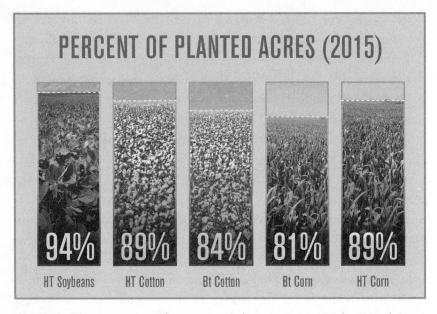

PERCENT OF PLANTED ACRES (2015)

94%	89%	84%	81%	89%
HT Soybeans	HT Cotton	Bt Cotton	Bt Corn	HT Corn

Figure 11. The proportion of corn, soy, and cotton grown in the United States that possesses either genes from bacteria that act as pesticides (Bt) or genes that allow the plant to tolerate herbicides (HT). Any given variety can have both; most do. *Data are drawn from the USDA Economic Research Service.*

any other country. It is often said that the rate of adoption of transgenic corn, soy, and cotton is "saturating" in the United States, leveling off. This is true, but only because the percentage of farmland occupied by transgenic crops as compared to their nontransgenic counterparts is approaching 100 percent.

· · ·

These new crops—initially corn, soy, and cotton with Bt and/or herbicide resistance—were genetically more homogeneous than perhaps any other crop ever planted at such a scale. The Bt crops (corn and cotton) were defended by a single toxin or, in later versions, only a few related toxins. If a corn pest evolved resistance to the most common Bt toxin, it would, subject to its ability to move region to region and deal with different climates, be able to devour much of the world's corn. The weeds were sprayed with huge quantities of a single herbicide. If a weed evolved resistance to the herbicide, it would grow between the corn, soy, cotton, and other crop plants (later crops with herbicide resistance would include canola, beets, and alfalfa), pushing up toward the sun like a religious zealot reaching for God. Fortunately, at least for Bt crops, there was a plan. It seemed that something had been learned from the fallout of the Green Revolution.

Using evolutionary theory, scientists developed an approach to delay resistance to Bt crops called refuge planting. If Bt corn, for example, were planted along with a crop that did not produce Bt, the non-Bt crop would serve as a kind of refuge in which pests susceptible to Bt could thrive. Every so often a pest would still evolve resistance to Bt. But when it did, the odds were that it would then breed with the far more abundant susceptible pests feeding on the refuge crop. Their children would then, having genes from both parents, not be sufficiently resistant to feed on the Bt crop.[9] The area of refuge necessary relative to the Bt crop could even be calculated based on mathematical models that are in turn based on evolution's basic rules. The amount of Bt toxins the crops need to produce could also be evaluated. Where this strategy was implemented, it works. Refuges are now mandatory in both the United States and Australia wherever Bt crops are planted. Elsewhere, the refuges are strongly suggested.[10] Here is the successful use of evolutionary theory to guide our future.

If resistance could really be stopped, Bt crops would be invincible. They and other modern genetically modified organisms would come to dominate globally, and the Red Queen of civilization would, once more, jump ahead. And, in practice, the advantage of these crops, the ability to reduce insecticide use and suppress pests, seemed to be paying off with both increased yield and higher farmer profits.[11] What's more, pests were becoming rarer, not only where the crops were planted but also on nearby fields. In addition, the natural enemies of crop pests, the beetles and parasitoids, the monsters often enlisted in biological control, were doing better when the new crops were sprayed with less insecticide.[12] An ecosystem had been engineered in which pests could be held at bay. Van Valen's hypothesis had found its limits. We were finally becoming a species able to determine its own fate, to postpone our own extinction. All we needed is for those people who planted Bt crops to follow the rules necessary to avoid resistance. They had to read, as it were, the label.

But as we've learned again and again in this book, the enemies always catch up. Resistant pests had started to appear. It happened first in the United States. A resistant caterpillar feeding on cotton bolls in a field, then more than one. Quickly many more than one. Then soon, resistant pests had been reported elsewhere on three continents. Five species are now known to be resistant to Bt crops, five and counting. For theoreticians, Van Valen's clan of clear-thinking outsiders, the good news was that the emergence of these resistant pests confirmed not just the indefatigableness of the Red Queen but also other theoretical expectations. The rate at which resistance evolved increased as the area of Bt crops increased. It increased as the number of pest species potentially exposed to the crops increased. It increased more quickly in those regions where farmers did not or were unable to plant refuge crops. This was an elegant confirmation that evolutionary biologists had a firm understanding of the dynamics of the Red Queen.

For the hundreds of millions of acres of fields at risk, there was nothing elegant about it. Once an insect evolved resistance in one place, it could eat its favorite crop anywhere. It was and is a problem with wings, a problem that continues to expand and has still not been resolved. One solution is to develop Bt crops with two or more toxins rather than just one, but these work best if planted before resistance emerges, because resistance to one toxin appears to increase the odds of resistance to a second. Optimal management of crops so as to reduce resistance requires a level of centralized agricultural planning and foresight currently possible in relatively few countries. Meanwhile, Bt crops, not to mention other related GMOs, have covered ever-larger parts of the earth. In 2015, the area planted in Bt crops was about one and a half times the area of the country of France, and growing, a country's worth of crops that, if managed just right, would for a long while remain impervious to pests. But when Bt crops are not managed just right, or pests do not meet the critical characteristics important for delaying resistance, pests can evolve resistance in as little as two years, no longer than it takes to evolve resistance to traditional pesticides.

If there is to be a long-term solution (no one believes, anymore, in permanent solutions), it will require better ways of predicting the evolution of resistance while keeping both the behavior of humans and the evolution of insects in mind. One model is for governments to encourage larger refuges through rewards of various sorts, though the trick in such an effort will remain that this will work best if all the countries that plant a particular crop, or at least all the countries in a region that do so, use a similar approach.[13] If one country fails to control resistance, the resulting pests are a risk to everyone.

But the other big piece in terms of the future of resistance is how the ever-larger agribusinesses interact with each other, how they behave. The leaders of these companies—emperors of the new agriculture—along with their chief scientists, wield enormous power over our global food supply. When these leaders gather in a boardroom in New York; Washington, DC; or Missouri, they are the most powerful people on earth in terms of what grows. They are the new Mother Earth. Their decisions influence what is planted across the empires where their seeds grow, empires as large as any that have existed in human history. These emperors are one other's enemies, yet inasmuch as their crops are similar (the number of versions of Bt genes is few, and the number of ways to achieve herbicide resistance is fewer), what threatens one of them threatens them all. As a result, they have an incentive to work together to combat resistance. Working together in this way is a sort of collusion, yet it is far easier for the companies to protect crops against resistant pests and weeds if they work together than it is if they are out of sync. But this collusion only works if none of the companies has a brand-new technology that the others lack. If one of the companies develops a new transgenic crop that kills pests, one that works far differently from the Bt crops, then the incentives of that company change dramatically. That company then begins to benefit if resistance to Bt emerges faster.[14] The emergence of resistance provides a market for the new crop. In such a scenario, little incentive exists for the company with the new technology to manage resistance well. It would be illegal for such a company to intentionally spread resistant pests, but anything short of that would probably be legal, legal and catastrophic for farmers and consumers around the world.

For now, though, this scenario—in which the company that is a little bit ahead in the Red Queen race sabotages the other companies— is not on the horizon. It is not on the horizon for the simple reason that none of the companies has yet figured out what to do if the worst pests become resistant to Bt crops. What happens then? This is a scenario that those who work with transgenic crops are already talking about in quiet conversations at conferences and meeting rooms, a scenario that few could have contemplated a decade ago. What happens when resistance of a devastating pest develops to all the Bt transgenic crops being used in a region and no new varieties are patented, ready to deploy? We may, in some regions, soon find out. As long as new transgenic crops are being engineered, such a collapse wouldn't be forever, but it could be long enough to make life very difficult in countries where most crop fields are planted with Bt plants. If the Bt crops all fail now, there appear to be no replacement transgenic crops in the pipeline for approval. Farmers would have to quickly return to using Green Revolution crops and pesticides at a massive scale or go back to using traditional seeds (if anyone, by then, has such seeds), and, to avoid major economic collapse, farmers would have to find a way to do so very quickly.

. . .

During the years in which Cary Fowler was working toward the seed treaty, he watched the early spread of transgenic crops and the continuing spread of Green Revolution crops with growing concern. He could nearly hear the *click, click, click* of Harlan's ratchet tightening the bolt that connected humanity's future to that of agroindustry. It was beginning to seem as if the quickest road to apocalypse was simply the failure of agribusinesses to keep up with the evolution of pests, pathogens, and climate change. He couldn't stop this trajectory, but he could do something else: he could create a seed bank that could be used in case everything went wrong, a seed bank for the apocalypse.

At first Fowler imagined a seed bank that could be used by anyone at any time. Then, when that idea proved intractable, he imagined a seed bank that could be used by anyone as a backup, an external hard drive of humanity's most beautiful creations.

Existing seed banks could share seeds among countries and breeders; they could fulfill the short-term needs. The global seed bank would ensure that seed varieties, all of them, could also be shared with children born a thousand years from now, ten thousand years from now, a hundred thousand years from now, regardless of how bad things might be in the next decade or century.

Still, as much as Fowler's vision was for a seed bank of the future, the biological realities of what such a seed bank would need to be were, well, humbly biological. Seeds are beautiful, potent, and sustaining, but they are also composed of living cells, living cells that must be protected from pathogens and their own willingness to metabolize and use energy. To be kept well, seeds must be moved into a kind of suspended animation. They must be kept dry or dried. Water triggers germination, rot, or both. They must be cold, which helps keep seeds from metabolizing. Then—and here is the big challenge—the seeds have to be tested every so often. Germinated. This can be hard. Some seeds require special signals to germinate. Smoke. Cold. Cold, then smoke. If they germinate poorly, they have to be grown anew.[15] Agribusinesses must do all this, but their methods, their costs, and even whether they are doing it at all are secret.

In the seed collection that Fowler imagined, all this was going to have to happen on a large scale, because the collection, like Vavilov's, was meant to be a collection for all of humanity. It was intended to survive wars, barbarous leaders, fires, floods, and each of a hundred other horrors of which nature and humanity are capable. If this were to work, the seeds would need to be consistently tended. The weakest link in the plan to save the seeds forever would be the humans needed to care for them, busy monks inside the cathedral of agriculture.

Helped along by the seed treaty, Fowler eventually found funding. Or, rather, funding found him. The Bill and Melinda Gates Foundation wrote asking Fowler if he could put $50 million dollars to good use in saving seeds. He said he could. He then found a location. The spot chosen was one of the northernmost inhabited places on earth, the northernmost place reachable by regular commercial flights—Svalbard, Norway. Svalbard is an inhospitable archipelago about the size in total land area of Scotland. It is too far north for trees and for most humans—except the 2,300 or so slightly crazy Norwegians who call it home (and a handful of Russians, thanks to an unusual old treaty). It is largely a landscape of reindeer, lichen, glacier, and the occasional polar bear. Fowler wasn't the first to think this place would be a good site; it was already the location of part of the collection (around ten thousand samples) of the Nordic Gene Bank.[16] At the site, the Nordic Gene Bank was (and is) studying the effects of storing seeds for a hundred years in the tundra on their potential to germinate in the future. Also, the Norwegians, rich on oil money, were willing to contribute the funds to build the facility. By 2004, final plans existed. In 2006, the first stone was laid.

The seed vault, once built, was to be called, perhaps not so creatively, the Svalbard Global Seed Vault (Svalbard Globale Frøhvelv, in Norwegian). Others, of a pithier and more dramatic nature, nicknamed it the doomsday vault. In Spanish, it is sometimes called Noah's Ark, except that the species being saved are escaping not a literal flood but rather a sea of industrial agriculture.

By 2007 the vault's construction had been started. By February of 2008, it was finished. It was not built so much as it was excavated, the chambers of the vault dynamited into a mountain. Most mines, like coal mines, dig into the layers of the past to burn it for energy. This was to be, Fowler thought, a mine into the future. The mine is high enough up on the mountain to avoid the risk from sea-level rise. It is well enough reinforced to be immune to earthquakes. And, though no one likes to mention it, it would survive the fallout from an atomic bomb. The seeds inside would survive the apocalypse, at least as long as a few botanists survived inside the collection with them.

Fowler would lead the work at the vault. Its finances were to be managed by an agreement between the Norwegian government, the Global Crop Diversity Trust, and the Nordic Genetic Resource Center. The Norwegian ministry of agriculture would pay for some of the operating costs. Donors including the Gates Foundation paid to support further seed collection.[17] They also paid, for example, for work done in partnership with seed donors (such as the USDA) to find seed varieties resistant to warm conditions and drought and to catalog the attributes of the seeds being put into the vault. But to do so, the donors would have to work with seeds they maintain themselves, since the seeds in the Svalbard vault are not to be used, not unless there is a real emergency.

In line with the policy on the ownership of seeds that Fowler helped draft, the seeds are organized according to the country or institute that provided the seeds. They are, in short, binned by the politics of the moment. The seeds of warring countries sit side by side but do not mix. Those who provide seeds to the seed bank pay nothing to store them. The collection's goal is to find all varieties of all crops, every single one. Fowler was, in his own telling, nervous the whole time the facility was being built. He was also nervous as he danced from seed collection to seed collection, trying to persuade people to share what they had worked so hard to save for themselves. But in 2010, just two years after it opened, the collection hit five hundred thousand species. Fowler wondered whether he might live to see the millionth variety enter the collection. If he didn't, he knew where he wanted to be. The team that built the vault left a small space in the wall where they said Fowler's ashes could go—as if his life were a votive, an offering to some very different future.

In 2012 Fowler stepped down from his role as director of the Global Crop Diversity Trust, leaving it and the seed collection in the hands of the current director, Marie Haga, and whoever might succeed her over the next thousands of years. Under Haga's leadership, the collection has continued to grow. By May 15, 2015, there were 864,000 varieties of seeds in storage. Fowler, to his own surprise, is close to seeing the day the collection hits a million varieties. It is not there yet, but it is nearly so—nearly a million varieties of seeds now safe from extinction, including most of the world's varieties of major crops, such as wheat and corn. And he feels healthy, too—healthy and accomplished, if not yet satisfied.

Vavilov would be pleased that the collection includes samples of most of the varieties of crops from each of his centers of origin. It also includes copies of nearly all the varieties of seeds from CGIAR institutes in those and other regions, including the International Potato Center, in Peru, and the International Center for Agricultural Research in the Dry Areas currently headquartered in Lebanon—particularly seeds from Mesopotamia. What it does not, unfortunately, include are most of the varieties from Vavilov's collection itself. The Vavilov Institute has provided just a thousand varieties out of tens of thousands, though no one knows how many of Vavilov's seeds are still viable. Clearly the work of guarding seed varieties for the future is not done. Many more seed varieties remain to be saved. A million more? Perhaps. The seed bank can hold between four and four and a half million samples. Many remain to be gathered, that much is evident, particularly from those regions that face climates like the ones temperate regions are likely to face in the future—climates in which few collections have been made because even seed collectors avoid the hottest, driest, most challenging places. The vault also does not include transgenic crops or crops from the major agroindustrial companies. It does not prohibit seeds from such companies (though Norway's regulations effectively do so). But inasmuch as such companies are focused on short-term profits (the next ten years at best, not the next ten thousand), they have little reason to bank their varieties of seeds. To the extent that agroindustrial companies comprise their own country, they form the biggest country not participating in the seed bank.

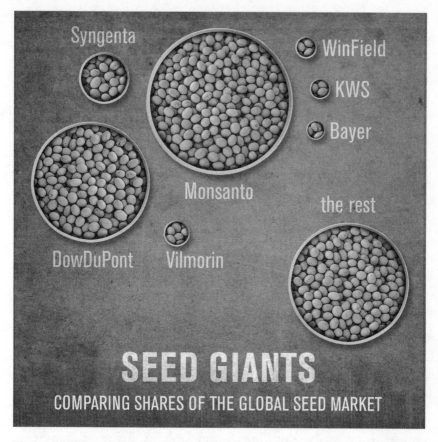

Figure 12. The market share of agroindustrial and other companies in seed sales, as of 2011. Most seeds in the world, especially those seeds that make up the majority of our diet, come from just three companies. *Data source: ETC Group.*

. . .

Agribusinesses do not participate in seed banks in part because their commitment is to the bottom line rather than to the future, and for the bottom line they do not need to rely on the seeds in public seed banks. They rely on the seed collections they purchase, combined with DNA from other organisms, to produce new transgenic varieties, particularly for economically beneficial crops such as corn and soybeans. Despite the seed treaty, these seeds, nearly all of them originally from centers of crop diversity, remain both secret and private. By contrast, research conducted by universities and other publicly funded institutions continues to rely on seed banks for breeding crops (and on Svalbard as a backup in case things really go wrong). In this light, a nearly inevitable future has begun to set itself up. It is a future in which transgenic crops move around the world in the same way the Green Revolution crops once had, like a great wave (the extent to which "great" is ironic depends on your perspective). We are getting nearly as much out of crop fields as seems possible, given the laws of physics and the rules of biology. In a world where food supplies need to double by 2050, transgenic crops are, some contend, the last magic bullet that will allow yields to increase, though the extent to which they have so far is, at best, debatable. There is the potential for reduced pesticide use, too, though this is perhaps a more fleeting possibility than one might have hoped.

To the extent that anything, apart from the evolution of resistance, has slowed the juggernaut of transgenic crops, it is the response of the public. The resistance of the public to transgenic crops has led many European countries to ban their use and/or import. It has shaped public policy and, with it, what is being grown where. The public has begun to resist transgenic crops for several reasons. Some resist them because of the risks such crops might pose to health. Yet to date, little evidence suggests any negative health consequences of existing transgenic crops.[18] Some resist transgenic crops because of their consequences for the environment, though again there is little evidence that transgenic crops are any better or worse for the environment than are their nontransgenic alternatives.[19] Some worry about the consequences we don't yet understand, which is always a reasonable concern with new technologies.[20] But none of these is the biggest problem with transgenic crops.

The greatest risk—and it is a big, big, one—posed by transgenic crops produced by industry is neither to our health nor to the environment but rather to our ability to continue to escape pests, pathogens, and changes in climate. So far, transgenic crops have accelerated the simplification of agriculture and the speed with which new pests and pathogens are evolving the ability to eat our simplified crops. Since the advent of transgenic crops, not only are we relying on relatively few crops for food, those crops are also, increasingly, defended by the same genes. The big worry as we look forward with regard to agroindustrial crops, including transgenic crops, is that they will continue to get less diverse. The danger lies in the extent to which the increasing shift to their use means that our ability to respond to agricultural challenges in the future is tied ever more to the ability of agroindustrial companies to respond.

The larger the proportion of our agriculture that is based on the crops for which the seeds are produced by the relatively few, immensely large agroindustrial companies (regardless of whether those seeds are transgenic), the larger the role these companies will have in responding to the movement and evolution of new pests and pathogens and climate change. Not only do agroindustrial companies have little incentive to diversify their crops, as long as our demands are so simple they also have little incentive to plan very far ahead. Just as in other industries, the purely economic incentives of agroindustrial companies will lead them to invest heavily in the few crops that sell the most and not to invest in other varieties unless they supply even more of the same. They have no incentive to plan for the next decade, much less the next century. They are encouraged only to provide more tomorrow, more of the same regardless of conditions, more of the same regardless of season. They give us the potentially most dangerous thing that any company possibly could, namely, what we ask for. Led by our demand, they give us food at our stores that is *never out of season.* In response to our simplest urges, they have given us food that is the same all year, a manifestation of our cravings. The preferences of our tongues and eyes have created the brave new world we live in. They have shaped what is in the store, shaped what is shipped, shaped what is in the field, and in doing so shaped the ecology of much of the world. The trouble is that this has also shaped the future, made the future ever more likely to be a place in which big agribusinesses save us just in time or fail to do so.

. . .

If agribusinesses do not help us flee the pests and pathogens of our crops or deal with climate change, we must have an alternative. Fortunately, the doomsday vault now exists, but it is an alternative of last resort, our go-to in case everything falls apart. We need more immediate and less apocalyptic alternatives. Historically we would have turned to the scientists at public institutions, scientists who can work with and learn from the varieties of crops in public seed banks. Such scientists continue to breed crops for new climates, crops that escape new pests and pathogens. The crop varieties scientists at public institutions breed might not be in our stores, but they would be ready, when times get tough, to rise to the challenge. Scientists at public institutions can look to the long term. They can anticipate the future rather than simply respond to the present. Historically, most of the responses to new pests, pathogens, and climates have resulted from the work of researchers funded by public institutions. In developed countries, these institutions were associated with universities. In developing countries they were associated with centers funded by the largess of developed countries. Many of these institutions were founded and funded in the wake of the potato famine. Such institutions paid the salaries of Norman Borlaug, Hans Herren, and Harry Evans, among others. The public institutions will save us!

The problem with this scenario is that the rise of the agroindustrial companies and the economic models that favor them has led to consistent decreases in funding for public research in plant breeding, plant pathology, entomology, and associated fields. It has led to decreases in the resources available to those who might, publicly, help in the race of the Red Queen. This diminishment has been rapid[21] and is associated, in the twenty-first century, with a slowing of progress in traditional breeding. Before 1960, researchers at public institutions were responsible for nearly all plant breeding apart from that being done by farmers. By 1990, twice as many plant breeders could be found in industry as in academia.[22] The less profitable a crop is for export, the worse the situation is. Cary Fowler estimates, for example, that there are fewer than ten yam, ten banana, and ten plantain breeders in the world.[23]

In the United States, the country that spends the most on agricultural research in public institutions, private-industry spending now dwarfs that of the federal government and universities. This is true for agricultural research in general but especially for crop breeding.

What's more, the starting point in the race is not even, inasmuch as industries are reluctant and unlikely to share their successes in crop breeding, so university researchers must often innovate from scratch. Where industries do share their insights with academia, they do so under nondisclosure agreements. They also do so in part as a means of employing cheap labor. Industry pays better than academia, so if Monsanto wants to test a new technology it is cheaper for the company to hire relatively low-paid professors than it is to pay its own employees to do the same task. Of course the research of the professors funded by industry, like that of industry itself, focuses disproportionately on crops that are of high value to industry in the short term.

While the funds available to institutions in the United States and Europe for crop breeding have decreased, the shift has been most extreme in tropical and subtropical countries, where crop research has historically relied on international support. Since Norman Borlaug began his shuttle-breeding program in Mexico, the program that jump-started the Green Revolution, agricultural scientists from around the world have gathered annually in Texcoco, Mexico, at the International Maize and Wheat Improvement Center (CIMMYT, or Centro Internacional de Mejoramiento de Maíz y Trigo). From Texcoco, they travel to Obregón, Mexico, re-creating a version of the trip Borlaug's wheat took as he raced it back and forth between the two sites to get in as many generations as possible per year. The journey is not mere nostalgia. As wheat breeders work on new varieties, varieties better able to deal with new climates and pests, they continue to employ shuttle breeding and, in doing so, retrace Borlaug's steps. Or they would, except that for the first time in twenty-five years, in 2003 too few funds were available to support the work. The shuttle breeding for the season was called off. Half the fields in Obregón sat fallow.[24] In 2004 the work resumed, but the message was clear: the work was not and is not valued.

The research at the two sites is now coordinated, just as it is at the International Potato Center, by CGIAR—which, in turn, is funded by the World Bank and its donor nations, who give less each year. For the purposes of comparison, Monsanto's research budget is nearly two hundred times what CGIAR receives for research from the World Bank. CGIAR works from Nepal to Bolivia; it is the biggest (and frankly only) such program in the world. All this is more problematic in light of the reality that as climate changes and human needs simultaneously become greater, we need to be breeding new kinds of crops, crops tolerant of drought and of new pathogens and pests.

Because of the declining support for publicly funded research on crops, the biggest worry for the future is not transgenic crops. It is not—or at least not to the extent that it was previously—the preservation of seed varieties. It is instead the possibility that our ability to deal with new challenges will fade, slowly, by attrition. This is the darkest ghost of Vavilov's tragedy. Vavilov's death was a tragedy. The deaths of those who worked to make his vision a reality were tragedies. At least, however, the seeds were saved. Yet in the years that followed, many of those seed varieties seem to have been lost to the ordinary ravages of budget cuts and time. That time and budgets remain a threat is obvious when considering the public research being done on crops. It is obvious, too, at Vavilov's Pavlovsk research station.

The Pavlovsk research station is the same station from which potatoes were rescued during the siege of Leningrad, rescued as the bombs fell. It contains the largest northern orchard of fruits and berries in Europe. Six thousand varieties of apples, a thousand varieties of strawberries, a thousand varieties of currants, and hundreds of other cold-tolerant nuts and fruits—including cherries and plums—grow there. Nearly all these varieties, perhaps more than 90 percent, including the thousand varieties of strawberries, can be found in no other collection. Many can no longer be found in farmers' fields, either. In the spring, the sight must be intoxicating, at least for the time being.[25]

The challenge for the station and its collection comes from the whims of affluence and the vagaries of law. A relatively new Russian law allows lands that are not being used "efficiently" to be taken over by developers. To developers, the station's land is the definition of useless inefficiency—trees with fruits and bushes with berries that are saved and studied rather than being consumed. Oil-rich Russians would like to build second homes where the orchards now stand. The station, after all, is an idyllic setting, near the palace of Catherine the Great and, of course, surrounded by an abundance of lovely trees and flowers. This story's theme is reminiscent of one of Anton Chekhov's plays, *The Cherry Orchard,* in which a family suffers the loss of their great house and orchards to industrialists. The last act in the destruction of the family's legacy is the cutting down of the cherry orchard.

In early 2011, government officials visited the orchards and pronounced them, officially, useless. The "plots are not being used," said Andrei Anisimov, deputy director general of the Federal Fund for Housing Construction Assistance.[26] On August 11, 2011, an arbitration court in Moscow agreed and ruled that the orchards should be given over to the Russian Housing Development Foundation to be turned into buildable lots and houses. The government informed the VIR staff that they would just have to move the plants.[27]

How hard would it be to move the plants? Impossible. To move them would be a task more herculean than running the orchard in the first place, which the institute only just barely has enough funds to do. Although the trees and bushes gathered by Vavilov's generation and cultured by those who followed remained, all that surrounded them was in disrepair. The real assault on the station had been the chronic lack of funding it faced, its decades of neglect. Moving the plants would be the end of the station; it would probably be the end of most varieties, too. Yet this seemed to be the orchard's fate. Clearing the land was to begin as soon as September.

Then, two days after the court decision, a reprieve. Under international pressure, Russian president Dmitry Medvedev tweeted that he would review the law that threatened Pavlovsk.[28] This review stalled things for a while. The trees lost their leaves and made it through winter. In spring, the orchard was full of bees and thousands of kinds of flowers. Then another year passed with no certainty. The bulldozers waited. Finally, on April 17, 2012, a federal order was written that allowed the orchards to remain, for the purpose of saving genetic resources.[29] The order is a stay of execution.

It comes with no more funds. It comes with the promise of more of what preceded it, slow neglect. This neglect is what makes the death of Vavilov and those inspired by him most tragic—they may have died in vain.

. . .

Cary Fowler, like many others, is dedicated to making sure Vavilov's work was not in vain. He fought for a version of Vavilov's vision in the form of the Svalbard seed collection. He helped fight to save Vavilov's orchard. He has tried to fight, only partially successfully, to save Vavilov's own seeds by guarding them in the Svalbard collection. Russian scientists and politicians have been reluctant to be part of the effort. If something happened to Vavilov's seed collections in Russia, most of the unique seed varieties therein would simply be lost.

Meanwhile, the seeds of Svalbard must be cataloged and studied. Because the seeds in Svalbard are a backup, this really means cataloging and studying the seeds in other seed banks—the ones that are backed up in Svalbard. Vavilov always imagined that this is what would happen with his seeds. He would grow each variety under multiple conditions and observe both its attributes (how fast it grows, how it tastes, how it defends itself against each of many pests and pathogens) and the extent to which those attributes differ depending upon growing conditions. This has yet to be done for most seed varieties in seed banks—nearly all of them.

In addition, to make use of these attributes, scientists need to study the genes of each variety, its full genome along with key genes and the variants associated with important traits, such as the ability to withstand a rust. Projects are under way that attempt to achieve these goals,[30] though they tend to focus on particular crop types, such as rice. We need to do this work for every kind of crop. Yet even the most hopeful agricultural scientist would agree that we are a long way from doing so. At our current rate, the effort will take centuries. No technical barriers exist to achieving this goal; it just costs more money than we are willing to spend. Meanwhile, something predictable is happening in public institutions where breeding goes on. Scientists at those institutions are moving away from traditional breeding and toward genetic engineering. They are doing so for the simple reason that it is quicker, and in an environment where funding is uncertain, it is a way to make an advance when there might not be funds to continue a long-term breeding trial the following year or the year after that. In fact, of the few new people who have been hired to work on plant breeding at universities, most focus on genetic engineering. This was true in 2014, but since then, something has happened that appears likely to dramatically hasten this trend. What happened was CRISPR.

CRISPR-Cas, where CRISPR stands for "clustered regularly interspaced short palindromic repeats," and "Cas" is short for CRISPR associated protein (evidence that scientists should just not be allowed to name things), is often described as a technology, and it is used as such. But the truth is that its innovation and elaboration were achieved not by scientists but by evolution. CRISPR-Cas is a system that evolved in bacteria and archaea. It allows bacteria and archaea to store DNA from viruses that attack them. They use their CRISPR-Cas systems to build libraries of old enemies. These systems allow bacteria, when confronted with an enemy, to compare that enemy to those in its library and, if there is a match, to cut up the DNA of the attacker. The CRISPR-Cas system is a means of precision defense. Once it was discovered, it became clear that it could be used for another goal, too. Components of the CRISPR-Cas system could be used to cut the DNA in other organisms at specific points and, having done so, edit it. In just a few years this technology has advanced so rapidly that it is now possible to edit the DNA of nearly any organism that has a sequenced genome and to do so with a precision never before imagined. CRISPR-Cas allows scientists to change single nucleotides of DNA in organisms—single letters in their genetic code.

Technology based on the CRISPR-Cas system offers a radical new step in the history of crop breeding. Ten thousand years ago crop breeding relied on natural reproduction among plants. Farmers then selected from among the resulting progeny. Two hundred years ago, crop breeders began to control which crop variety or species mated with which and began to intentionally produce progeny with specific desired traits. A hundred years ago, breeders began to cross inbred lines in order to breed specific traits of one variety (or even species) into another high-yielding variety. More recently, transgenic technologies broke the barrier of who could reproduce with whom. CRISPR-Cas may be potentially a bigger shift than any of those that preceded it. It allows scientists to copy individual genes from one species and insert them into another species with high precision, not only in terms of which genes are moved but also in terms of where they go. CRISPR-Cas is relatively fast, too. Traditional crop breeding typically requires seven to ten years to produce a new variety (and even longer for trees); traditional transgenic approaches require five to seven years. Using CRISPR-Cas, producing a new variety takes as little as three years and is getting quicker all the time.

It is important to note that, unlike transgenic approaches to producing new varieties of crops, CRISPR-Cas can be used to alter the genetic code of organisms in ways that are undetectable and are likely to remain undetectable. Whereas traditional transgenic crops always include the DNA of a foreign organism, typically more than one, organisms altered with CRISPR-Cas do not need to. Instead, technologies based on CRISPR-Cas edit the existing genome. This may seem like a subtle difference, the difference between cutting into a slab of clay as opposed to adding extra clay to the slab, but it has many effects. For one, it means that most of the existing regulations on genetically modified organisms, regulations focused on the insertion of DNA (rather than alteration of existing DNA), are not applicable to plants or other organisms modified using CRISPR-Cas. For another, it means that even if new regulations for organisms altered using CRISPR-Cas emerge, enforcing them will come to depend on the honesty of the agroindustrial companies. Honest self-reporting as a means for enforcing policy is unlikely to yield strong compliance, particularly when there are billions of dollars to be made.

What's more, using CRISPR-Cas is cheap (again, relatively speaking). Every force for good in the world can use it, but so can every force for evil. North Korea may struggle with nuclear weapons, but it should have far less trouble using CRISPR-Cas to produce agents of agricultural terrorism, for instance. A new world has been cracked open. In this world, the limits to producing new crops will be the availability of crop varieties, the availability of wild species, the insights as to what to mix with what, and the funds to bring new crops through each of the necessary regulatory steps. Ironically, the more CRISPR-Cas and transgenic crops in general are regulated, the more dominant agroindustrial companies will be, because those companies will be the ones able to pay for each of the steps necessary to bring a new crop to market. In one scenario, new crops may well become like new antibiotics—so expensive to produce that the barrier to producing new ones is not the science but the steps between the science and the market. Or, in another scenario, new crops may become so easy to produce using CRISPR-Cas that regulation becomes impossible and the market is flooded with crop varieties.

Yet again, when you have a new hammer everything looks like a protruding nail. CRISPR-Cas is the most revolutionary hammer yet to be wielded by biologists. It may well change the ways in which we treat congenital diseases, breed new crops, and even produce antibiotics. It may also create problems of a scope we don't yet imagine. Which scenario is more likely remains unclear. Reasonable folks disagree. What is clear is that even if we are to focus on the use of CRISPR-Cas to breed new plants, we will still need to save and understand the crop varieties in seed banks. We will still need to save wild species, be they plants, animals, or bacteria such as those whose study led to the discovery of CRISPR-Cas.

CRISPR-Cas makes the genes of potentially any species on earth valuable, so suddenly we need to conserve not only the wild relatives of crops for agriculture but also all other wild species, however many they might be, wherever they might be. The dominance of agroindustrial companies makes our agricultural future dependent on their decisions and innovation. The new dominance of CRISPR-Cas makes their decision and innovation contingent on our ability to save and understand all of wild nature.

. . .

You have agency in this story, as the philosophers would say. You have control over the future of agriculture. You can vote to fund resources for those who breed crops. You can encourage your elected officials to support investment in international projects that aim to fund public plant breeding, especially of crops in tropical regions. You can, as Cary Fowler did, advocate on behalf of farmers around the world. But your other method of influence is your purchases. The idea that your purchases can help influence farming is not a new one (it is, after all, the idea that Fowler helped articulate in *Food First*). But it is no less true for being time-honored.

You can buy diverse varieties of local crops. Some scholars have even called for a massive return to locally grown food, a return of the urban masses to the fields that each of us might not only farm, but farm many, many varieties of crops using agroecological approaches. For many reasons, it is hard to imagine altering the agricultural world to such a massive extent (nor is it clear that we would want to). Fortunately, we can alter the diversity of crops without changing how the average bite of food is produced. We can affect the margins, and in agriculture a great deal has always depended on the margins. By increasing the proportion of food that is purchased from locally grown and diverse varieties of crops we increase the incentives farmers have to plant those varieties. We increase the incentives farmers have to find unusual varieties. And, importantly, we increase the willingness of farmers to experiment, whether with unusual crop varieties or even with the breeding of novel crop varieties. All of these benefits accelerate if government policies (be they local or national) support the ability of farmers to diversify their crops and focus on local varieties. All of this may seem like some hippy dream, but the truth is that in parts of North America and Europe as well as elsewhere local food movements have already increased the diversity of crop varieties available in seed catalogues and stores. For the first time in a hundred years, the total diversity of crop varieties available is increasing, even if the average bite of food is less and less diverse. This is not a panacea, and yet it is a move in the right direction, a move we can help to facilitate with each purchase, with each bite. Where we can, we need to encourage the culture of local, diverse eating, a culture that values the richness of every bite, a richness that we can savor, but that will also, ultimately, help us to sustain ourselves.

The title of this book reflects what happens when our sensory biases are coupled with industrial crop production. Our taste preferences have shaped industry, shaped the world, shaped what is farmed where. Our taste buds have led us to prefer whatever crop most cheaply provides sugars or fat, regardless of how or where they grow. Favoring—cultivating, even—a culture in which we consume more locally grown foods, and foods of more varieties, may help save our crops and, with them, our civilization from destruction. It may help save us from famine.

Perhaps it is too dramatic to suggest that what you eat and buy helps save us from apocalypse, saves us from having to use the Svalbard seed bank. Maybe a world in which our food is homogeneous and without season is okay. After all, right now, we still have plenty of food. The United States, for example, has large food surpluses that are wasted. We also waste huge quantities of our crops that we feed to domesticated animals, food we might eat ourselves and, in doing so, become much more efficient and sustainable users of the planet. One might also note that there are areas we could farm that we haven't—for example, our yards. What's more, the cynic might argue, even if food shortages do occur, they are unlikely to occur in developed countries. They will instead occur in places such as Central America, North Africa, and the Middle East. Yet what the Irish potato famine teaches us is that the loss of a crop in one region has consequences for the entire world. Nowhere is this clearer than in Syria.

15

Grains, Guns, and Desertification

For the first time since cities were built and founded,
The great agricultural tracts produced no grain,
The inundated tracts produced no fish,
The irrigated orchards produced neither syrup nor wine,
The gathered clouds did not rain, the masgurum tree did not
grow.
— "The Curse of Akkad," ca. 2100 BCE

A fter the US invasion of Iraq in 2003, archaeologists rushed to save the country's antiquities. Saddam Hussein's regime had begun to fall. The chaos of the transition threatened the country's treasures. Scholars gathered much of the art and many of the antiquities from Iraq's national archaeological museum. In doing so, they faced many of the same challenges faced by those attempting to save the art of the Hermitage when Hitler surrounded Leningrad. Unlike the objects in the Hermitage, though, many of the great antiquities of Iraq were lost or destroyed. Among the most significant treasures of Iraq were seeds.

The seeds of Iraq are the great works of art and science produced by the traditional farmers of Mesopotamia. Each, in its genes, bears the handprints of antiquity, the labor of thousands. Iraq's seeds are the direct descendants of the seeds on which Western civilization was built, descendants that have been improved over ten thousand years of farming, winnowing, and exchanging. Before the wars began, Iraq was self-sustaining with regard to food. If it was ever going to be again, it would need its ancient seeds. The very seeds threatened by the war. Fortunately, at least some had already been duplicated in other collections.[1] But many existed only in Iraq, whether in the field or in seed banks. Those in farmers' fields, if they were to survive, would have to endure the atrocities of the war alongside the farmers. The fragile seeds were tied to the equally fragile lives of women and men. The seeds in the seed bank would just have to get lucky, but given their location, the odds of that seemed slim. The seed bank of Iraq was stored in Abu Ghraib, a city just west of Baghdad bombed and attacked by both sides (it would become famous as the site of a US prison). Fortunately, those who worked to bank Iraq's seeds did not trust luck.

The seed collection from Abu Ghraib was gathered up in 2003. With little announcement, the seeds of antiquity, the seeds of Mesopotamia, the seeds directly descended from those that formed the basis of the first real civilization, were packed into a cardboard box. The box, after being folded and taped shut, was sent to ICARDA, in Syria—the International Center for Agricultural Research in the Dry Areas. There the box was put on a shelf for storage, the unceremonious fate of Mesopotamia's most direct legacy. Back in Iraq, the building that had housed the seed collection was bombed. By the United States, perhaps, or the British. The seeds had been shipped just in time. They could be used in Iraq again, after the war,[2] whenever that time might come,[3] used again in the fields where they and Mesopotamia arose. In the meantime, the US government, in an attempt to restore agriculture, had seeds from American companies distributed in Iraq in 2004. It was part of Operation Amber Waves, which included Order 81,[4] prohibiting the seeds the Iraqis were given from being reused. They would have to purchase the seeds—based on plants domesticated in Mesopotamia—from the American companies each year.

. . .

We used to describe what happened somewhere in or near Mesopotamia more than ten thousand years ago as if it were an invention. We picture a gatherer in the high grass. She is carrying a basket filled with hard-won grains. Tired, she is struck in an instant by a bolt of insight. Farming! She can plant the grains and grow them!

More likely farming was understood to be possible in one form or another long before it was practiced with regularity. For the grain gatherers, how hard would it have been, when a few stored grains sprouted leaves, to set them in the soil? How hard would it have been to sow a few? Yet for most of the million years of human history in Mesopotamia and the broader Fertile Crescent, no one did.

Then circumstances changed. Beginning around 10,000 BCE, people settled in small villages in the Zagros mountains to the east of Mesopotamia (today's Iran) and began to gather the seeds of wild grains, chickpeas, and lentils. With time, the seeds of these plants were planted. Why this might happen is an open question. Perhaps more food was needed. Seeds were sown as a way to buffer the lack of food in times of scarcity—perhaps the scarcity imposed by the confluence of population growth and changes in climate.[5] What we do know is that eventually grain seeds were planted. Once they were, each new year farmers wrested a little more control over what was being grown. At the end of each year, the seeds of the plants that grew better—the ones that had better flavor, had seeds that didn't fall out (shatter) when simply nudged, produced larger seeds, matured early, and otherwise suited the needs of the time—were more likely than other seeds to get stored for the next year.[6] Those varieties that were no good were burned or pulled, disfavored in some way or other. This cycle was repeated generation after generation and, in the process, wild plants were domesticated. Each story of domestication was unique. And yet the general features of the story were the same everywhere.

With the very first plant domestication—wheat, barley, chickpeas, and lentils—came larger settlements and then, ultimately (around 3500 BCE), the first civilization, Mesopotamia, in the region between the Tigris and the Euphrates Rivers. It started in the wetter southern region (in what is now Iraq) but would ultimately spread north as well (into what is now Syria). The first wheat was durum wheat (*Triticum durum*),[7] what we now call pasta wheat, to be followed by common wheat, or bread wheat (*Triticum aestivum*).[8]

In this same region, in about 2350 BCE, writing began. In Mesopotamia, empire began when, in 2334 BCE, Sargon of Akkad subju-
gated the cities of southern and northern Mesopotamia under his
rule.[9]

Amid the modern realities of the region it is easy to forget that
Western culture and food emerged in the Middle East. You might
not be genetically from the Middle East, but many features of your
daily life are. Mesopotamia was the beginning of the West. But
Mesopotamia was more than a beginning; it was also a kind of end.
Once humans started to farm, they were wed to the field. There was
no way to return to gathering. Populations grew fast, thanks to agriculture, and as they did they became ever more dependent on their
crops. The crops, in turn, became ever more dependent on people.
When the people suffered, the crops suffered. When the crops suffered, the people suffered. Farming was not an insight; it was instead
a marriage, a bond between humans and seeds. It is the marriage on
which the subsequent civilizations of the Old World, be they Persia,
Greece, or Rome, are based.

This bond[10] was forged thousands of years ago; it is a bond from
which we cannot escape. We are now charged with spreading the
seed of plants forever, as long as our kind might hope to exist. We
hold back pests. We provide fertilizer and water. The plants, they
give us sustenance in return—food made from sunlight, carbon
dioxide, and the minerals hidden between the grains of desert sand.
Those who study farming study ways to ensure that we get as much
as possible from plants.

Ahmed Amri is one of those who now works to save the agriculture of Mesopotamia. Amri grew up at the edge of a great desert.
On bad days, sands from the Sahara darkened the sky. On good
days, it rained, and his region was one of the few places in the country where water flowed and the landscape grew green with leaves.
As a boy, Amri became interested in the farms at the edge of the
desert and how to improve them. He wanted to become, if not quite
a farmer, someone who might make farming better, easier, or both.

Amri finished his schooling, up to his undergraduate degree, in
Morocco. He then signed up as a graduate student in the department

of agronomy at Kansas State University. Once there, he studied which of the wheat varieties from the Middle East were resistant to the larvae of Hessian flies, a pest that attacks wheat stems. The Hessian fly has spread everywhere wheat grows; it is diminutive and devastating. It poses the biggest problem in places where farms are small and pesticides expensive, such as northern Africa and central Asia. The goal of Amri's work and that of his colleagues at Kansas was to breed varieties of wheat resistant to the Hessian fly. Amri made good headway. He even identified the individual genes in some wheat varieties that seemed to be responsible for resistance.[11] It was beautiful work, elegant in a way that academics like—but it was not yet useful. However, the skills he learned in doing this work were useful, and they could be applied right away. Amri chose to apply them back in Morocco, where he worked to breed new varieties of wheat and barley tolerant to drought and resistant to pests. His was the hard, necessary work of counting seeds, making crosses, waiting for plants to grow.

While Amri was working in Morocco, another job came up, this one in Aleppo, Syria, at ICARDA. Amri took it. At ICARDA Amri worked to coordinate the genetic resources of the center, its collection of seeds, roots, and tubers.[12] Aleppo, often said to be the oldest inhabited city in Syria, was chosen as the new site for ICARDA in 1980 because Syria was stable and prosperous. It had the right climate for research on arid lands and is located in the heart of the Fertile Crescent, the arc of rain-fed agriculture lands that surrounds ancient Mesopotamia. Also, Beirut, the city in which ICARDA was first founded, in 1977, had become an active war zone.

By the time Amri arrived, more than a hundred employees of ICARDA were already working to improve the farming of crops and livestock that do well in dry areas. In the manner of Vavilov, they stored seeds. ICARDA's beautifully curated collection held more than 141,000 different seed samples. The collection is one of the largest of barley, beans, chickpeas, and lentils, all of which were domesticated in or near Mesopotamia. But it is especially impressive for its wheat, which is represented by 38,000 varieties; barley, by 29,000 varieties; and the wild relatives of wheat, by 8,000 varieties.

Beyond storing seeds, though, ICARDA works with farmers to use those seeds. From the farmers, they learned what the worst problems in fields were. They also learned about other varieties of seeds not in the collection. They then systematically tried to improve both the seed collections and the crops. To improve the seed collections, they predicted where in the region local crop varieties would be most likely to have the ability to adapt to conditions (drought, heat, salinity) and resist pests and went there to search for new seed varieties. They then used every trick in the agricultural book to breed new crops. And it all worked. Wheat varieties have been bred that are resistant to sunn pests (kin to the stink bug),[13] some Hessian flies, and powdery mildew. And new varieties have been bred that grow better under conditions of drought and high salinity. ICARDA was and is, in short, dedicated to the same things to which Amri is dedicated, at a large scale in the region in which, in many ways, they most matter.

ICARDA is important to all of North Africa and the Middle East. This entire region is dry and dependent on its own agriculture and agricultural innovation and always has been. ICARDA is not the only international center for crop breeding in the region, but it is the largest. Most others tend to be focused on particular countries, and in a region divided deeply by politics, breeding and agricultural centers focused on particular countries do not always help all mouths equally. In addition, the region, although dry, differs in terms of how dry and in terms of what sorts of crops are farmed. North Africa and the Middle East together are now thought to include not one but two centers of crop diversity, centers that layer on top of ancient history and also are embedded in modern struggles.

Though Mesopotamia and the Fertile Crescent were historically fertile, both in terms of their soil and their agricultural innovation, modern farmers there struggle. The fertility of the region has always been bound to the rivers and the rains—bound to faraway rainfall, local rainfall, and the ability to irrigate. Once, this water and the crops it sustained were enough for the people of the region. Today, however, while small farmers plant a great richness of varieties, the total quantity of food produced in the region is less than is needed by the people who live there. On global maps of agricultural productivity, the pockets of farms in the region are so sparse they don't even show up. The condition of the North African and Middle Eastern centers of crop diversity is like that of other such centers around the world.

Since the return of the conquistadors from the Americas, the productive agricultural regions have become those in relatively cool, relatively northern countries. In those countries, the affluence that followed the era of the conquistadors led to greater investment in science, agricultural intensification, and, ultimately, crop yield, which in turn led to even more affluence. This cycle was self-fulfilling. Ever greater intensification of agriculture led crops to be produced more cheaply, too, which reduced the value of the same crops when planted by small landholders (in the tropics and North Africa, for example). With the value of wheat too low to yield a profit for a small farm, agriculturalists in these regions shifted to cash crops for export at the expense of their own food. North Africa and the Middle East are like much of the tropical and subtropical world in terms of their place in the geographic push and pull of crops. In terms of climate, however, they are unique.

While models of the future climate of some regions of earth— for example, the southeastern United States—are uncertain (some models predict the Southeast will get wetter, others that it will get drier), little ambiguity exists in models for North Africa and most of the Middle East. All predict they will become much, much hotter and much drier, receiving half the rain they currently receive. A study conducted by Pinhas Alpert at Tel Aviv University predicts that the Euphrates River, lifeblood of civilization, will be 25 to 70 percent drier by the year 2100, even before considering any changes in the use of its waters.[14] This would not be the first time drought has struck the civilizations of Mesopotamia and the rest of the Fertile Crescent.

Mesopotamia was once ruled by leaders in a group of cities on the border between what are now Iraq and Syria, cities strung like jewels along the rivers. These cities were agricultural cities that, under the rule of the grandson of Sargon of Akkad, were united into an empire (just how united is the subject of the sort of endless debate that will keep archaeologists passively-aggressively furious and engaged at meetings for the next century). Each city held tens—in some cases hundreds—of thousands of people. These cities had their roots in the first villages in the region where crops were domesticated, but from those simple roots they grew.

Then, in roughly 2200 BCE, everything changed. The cities collapsed. In reconstructing this collapse, one reported mostly in story rather than in artifact, historians struggled to discern truth from fable. They struggled, even, to know for sure if there had been a collapse. Maybe, they said, the myth was just a lament, an ordinary sort of whining, about the empire's challenges. Maybe it was a moralizing myth that stood as an example not of something that had happened but of what not to do. It could not have been a true story of catastrophe.[15] Then the discovery of the city of Tell Leilan by archaeologist Harvey Weiss, on the Khabur plains, in Syria, began to clarify things. Tell Leilan was in the northern half of Mesopotamia, a region that had been, until Weiss began his work, relatively poorly studied in comparison to the southern parts of the region, in Iraq.

Harvey Weiss and his colleagues excavated Tell Leilan beginning in 1979. The occupation of the site was so ancient, going back to at least 6000 BCE, that Weiss and his team needed to dig their excavation trenches six meters (eighteen feet) into the ground to capture the whole story, a story on top of which eighteen feet of sand and time had settled. In those trenches, they found a clear chronology of the city. They saw, in the layers of walls, broken bricks, and dust, the markers of a growing population, measured by the geographic size of the city. For thousands of years, the settlement was the size of a village, but bit by bit, millennium by millennium, it grew. By 2800 BCE, it was thirty-seven acres. By 2400 BCE, it was two hundred acres. Then in subsequent years, once the grandson of Sargon conquered the city and incorporated it into the empire of Akkad, the city grew even more, as did the scale of agriculture. Wheat was grown; barley, too, as well as olives, grapes, peas, sweet peas, chickpeas, dates, and safflower (for dye).[16] The city had a wide, paved street and planned settlements where people drank beer and wine and ate bread and the meat of goats and pigs. Durum wheat, emmer wheat, barley, and other cereals were stored in centralized government facilities and distributed by the government in standardized sealed containers, marked with the city's logos, administrative seals, all of them showing banquet scenes. It was a major city within a broader agricultural empire, and it was growing fast: then, all of a sudden, in the strata of the excavation pits closer to the surface, the evidence of the city and its occupants, after four thousand years of habitation, disappeared. As Weiss and his team dug through the layers of time, the layer corresponding to 2200 BCE seemed to be empty. No evidence of humans. No domestic animals. Not even the evidence of earthworms could be found. What's more, the layer in the trenches they dug corresponding to 2200 BCE and thereafter seemed to be filled with dust. It was as though the people had just disappeared. This is just what Weiss thinks happened.

In 1993, Weiss suggested that the layer of sand devoid of human artifacts was a sign of a drought, of hunger, and the consequent death and emigration of the people of the first great empire ever to have existed.[17] The cultural evidence of emigration was found elsewhere in the region, Weiss noted; in southern Mesopotamia, clay tablets mention the arrival of refugees from northern Mesopotamia. Similar transitions (though of varying extremes) were being discovered in archaeological sites in the Aegean, Egyptian, and Indus regions.

Climatologists were skeptical. But when they began to look for evidence of drought, they found it. First they found ancient evidence of links between the climate of the eastern Mediterranean and the flow of the rivers of Mesopotamia.[18] This was suggestive, but doubt persisted. Then they noted that during the period in which Mesopotamian cities collapsed, a drought had occurred in the eastern Mediterranean. The eastern Mediterranean dried, and, because rains come from the Mediterranean to Mesopotamia (or fail to), so, too, did the Tigris and Euphrates; so, too, did the agricultural fields alongside them. And with the collapse of those fields occurred the collapse of the cities they sustained. More recent research suggests that just prior to the collapse, in the moments when food was becoming scarce, the health of those who remained deteriorated. Infections became common, bones became stunted and weak, and anemia was widespread.[19] Then there was one final piece of evidence. Cities did not reappear in the region until the drought was over, nearly three hundred years later. Here, then, was a terrifying model for the future, especially since future climate models are clear that a drought will come again to the region and that it will never, not for hundreds or even thousands of years, be over.[20]

In light of the predictions of long-term regional drought (and the foreshadowing provided by the region's past), finding and saving all the crop varieties and wild relatives of crops able to deal with dry conditions, and breeding those crops to deal both with drought and

Figure 13. Relative expenditure on crop research and other efforts by ICARDA in 2006 compared to the *weekly* budget of the US government for the war in Iraq during the same period. *Data source: ICARDA annual report.*

heat, is key to the survival of hundreds of millions of people in the future. It is key, too, to the political stability of the region. Governments fall when their food supply collapses. ICARDA and other research centers must find new ways to feed millions of people on an ever-decreasing supply of water. They must do so with relatively modest funding, despite the significance of their goal and despite the comparatively huge cost of the alternative—war.

.   .   .

For Ahmed Amri, once he had moved to ICARDA, some of the aspects of the work were similar to his experience in Morocco. But at ICARDA he had and felt more responsibility to the center as well as to the region. Because ICARDA is *the* center for agriculture in dry regions, the team with which Amri would work there was far larger and the collections of seeds far more expansive than they were in Morocco. In addition, with wars in countries near Syria, it was a hub of peace on which those other regions, including Iraq, would ultimately depend for the future of their agriculture. ICARDA was the "lender of last resort"[21] for countries in need of seeds after war. It was also the go-to institution when seeds needed to be saved before war. So predictable were the wars of the broader region that this pair of needs had been identified and fulfilled repeatedly. When Afghanistan's national seed bank was looted, for example, in 2002, it was ICARDA that sent seeds so that crops might be planted the following year. Amri was present for that shipment. Then, just a year later, came the war in Iraq. Amri was present for that shipment— that cardboard box of seeds—as well.

As of 2005, the Iraqi seeds stayed on the shelf at ICARDA. Amri curated them. He spent a lot of time thinking about the ways in which seeds can save or fail to save a country. Of course, it would not be Syria that needed saving next. Not in a region that included Iraq to the east, Jordan to the south, Israel and Egypt to the west, and troubled Yemen not so very far to the south. Syria was the stable place, the place Iraqis and Afghans sent their seeds for safekeeping. Syria was the logical place to save the seeds, because among countries of the Middle East it was where things worked. Doctors were trained. Scholars were trained. Nearly everyone was educated. It was, by most met-rics, akin to the United States in levels of education and affluence; better off, even, than the state in which I live, North Carolina. This is worth remembering as we consider what happened next.

Beginning in the winter of 2006, Syria and Iraq, along with the rest of the Fertile Crescent, began to suffer a persistent drought. It was among the worst in the region's well-chronicled modern history. The drought affected the region in general, but especially the parts of Syria and Iraq that most directly correspond to the core of the ancient Mesopotamia. It came after decades of pressure from the government of Syria under president Hafez al-Assad to focus on farming that relied, to a greater extent than at any moment in its long history, on irrigation, using the country's scarce groundwater supplies. This drought was like previous droughts, only more intense (and, as it would turn out, longer). Because it was coupled with droughts in the headwaters of the Tigris, in Turkey (and, by some accounts, exacerbated by reductions in the amount of water that was released downstream by Turkey), water levels reaching Syria were low, as was the amount of rain falling from the sky. Because the drought came after the intensification of agriculture, which made access to groundwater supplies more difficult, little in the way of other sources of reprieve existed.[22]

As early as 2006, the very first year of the drought, international scholars began to worry a little about Syria and about the country's seeds in particular. They worried about whether the seeds needed to be moved. An editorial in the journal *Nature* suggested it would be better to give more funding to ICARDA and to other seed banks rather than to move the seeds. The United States was, at the time, spending $1 billion per week on the war in Iraq, and, the editorial argued, a $260 million endowment would be enough to save seeds from around the world for good, particularly if it was given to the CGIAR consortium office,[23] the umbrella organization that connects ICARDA to similar centers. Even if one added a zero to such a contribution it would still just repre-sent, at $2.6 billion, two weeks of war funding and less than one-tenth of the annual value of crop losses resulting from pests and pathogens in the United States alone. Such an effort would be of value not only to

Syria but also to the nearly 1,750 seed banks and their associated 7.5 million seed samples around the world, all but forty or so of which are struggling to meet international standards of long-term seed saving— struggling even as climate change threatens crops most in the very regions where these seed banks are most diverse. All these things were written about and said, but of course no new investment came. Also, the conditions in Syria got worse.

Crops began to fail. The biggest problem was wheat. Up until 2007, the production of wheat had been increasing in Syria without a corresponding increase in the amount of land being farmed. Between 1991 and 2004 alone, wheat yields doubled. These increases were attributable to the Green Revolution crops (and irrigation of those crops) and then subsequent work on ICARDA's part to breed new crops. There were bad years, but the net trend was an increase, largely in step with population growth. Then came the drought.

In 2008, wheat production declined by 38 percent relative to 2007.[24] In theory, if someone had acted fast enough, the great diversity of Syrian traditional varieties and the new varieties being bred by ICARDA might have been able to deal with the drought. If not in Syria, the site of ancient Mesopotamia and modern ICARDA, then where? In practice, though, whatever dry-adapted crops were present were too few and far between, or simply unable to help fast enough, so villages and their crops dried up, puckered from the heat. While many crops can grow well in arid conditions, few grow well when they must be planted during a drought. Agricultural production declined. More important, the breeding of new crops does not occur fast enough, even with modern techniques, to deal with problems once a crisis has arrived. It must be done in advance.

By the end 2010, Syria was precariously close to having to import wheat for food,[25] to import the very crop domesticated in the region. Children suffered from diseases as a result of malnutrition. The problem was made more difficult because the population had grown and become more urban. The urban population increased by roughly 50 percent between 2000 and 2010. This growth included a million Iraqis who moved to Syrian cities in response to the wars, but also 1.5 million Syrians who moved to cities from rural areas (in part due to the failure of crops). Once in the cities, people had few opportunities. The number of people vastly outstripped the number of available jobs. Such rapid change heightened existing instability in the cities, in the country, and in the region. Government policies under the new president, Bashar al-Assad, made these challenges more acute rather than easing them.

Two more hard years passed in 2009 and 2010, springs in which little grew. These years of drought, climate scientists have said, are in line with what is expected to occur more frequently in the region because of the influence of global climate change. Such drought is predicted to become more common and extreme in the coming decades. The years of drought were, in short, a harbinger. The year 2011 once more began as a dry year. Wheat, for the first time in decades, had to be imported. Then, in the dusty, desperate spring, a spring in which few seeds germinated, revolution bloomed.

Historians will long debate the exact dynamics that triggered the uprising. Hopeful citizens revolted against the oppressive Assad government. The government cracked down with great and horrible force. The revolution turned into a war, and in the unrest of this civil war, ISIS—or Daesh, as the group is known in most Arabic-speaking countries—began to seize power in parts of the country, the parts most closely allied with ancient Mesopotamia. The revolution, along with the rise of ISIS, triggered the collapse of the country, the deaths of several hundred thousand Syrians,[26] and an immigration wave larger than any since the Irish potato famine— by some measures even larger than the potato famine exodus. No fewer than four million Syrians have left the country: families pack into boats, ride on the backs of trucks, stack themselves into the backs of trains, and flee in any way they can.

. . .

In the months when the situation in Syria was getting worse but not yet apocalyptic, even before the uprising, Ahmed Amri decided he needed to get ICARDA's seeds out of Syria. The ICARDA facility, including its seed bank, had generators but not much fuel. It would not take much for the seeds to be destroyed. Like all those who work in seed banks, Amri knew the stories of other seed banks that no longer exist, that were destroyed. But unlike the other seed banks, ICARDA's bank was relatively prepared for tragedy. It had already sent duplicates of 87 percent of the seed samples in the collection to other collections, to be saved "just in case." The duplicates would prevent the total loss of those varieties, though duplicate copies of seeds are typically kept in small numbers, such that if they are ever to be used again, they would have to be grown out in order to produce more seeds. But the rest of the collection, which was huge—more than fourteen thousand samples of who could be sure how many varieties—was not backed up. Amri decided that those seeds, those samples, needed to leave.

The seed-saving mission began in the spring of 2011. It was horrifyingly similar to the efforts of Vavilov's team. The seeds were taken, by truck, to the north, away from the war, toward the border with Turkey. At the Turkish border, the Turkish representative of the ministry of agriculture met the seeds and helped clear them rapidly through customs and transfer them to the gene bank in Ankara. Other seeds went to Lebanon.

Eventually nearly all the non-Syrians at the center left. They were advised to. This included essentially all the scientists. No one could blame them. Amri stayed until the first week of July in 2012. Then he went back to Morocco, where he would continue to work on behalf of ICARDA, at a distance from the Syrian center.[27] His Syrian colleagues stayed guarding the seeds, much as Vavilov's colleagues did in Leningrad. As in Leningrad, though, the value wasn't just in the seeds. Around the ICARDA buildings are fields where the varieties of plants with the most promise as crops for dry regions are being grown. As late as May of 2014, the few Syrians left at the seed bank in Syria continued to guard the seeds. Then the war arrived at their door.

Almost any knock on the door would have been bad. In Syria and Iraq, ISIS destroyed several of the most treasured archaeological sites, including ancient Nimrud, one of the oldest cities in the world. ISIS was imitating the actions of those who colonized the ancient cities of the Akkad empire after its fall. When drought struck, the Gutian peoples came over the hill and overran the Akkadian homeland in southern Mesopotamia. They ignored agriculture. They released the animals. They captured women and children (at least in the telling of the Summerians, who also said that the Gutian people had human faces, the cunning of dogs, and the bodies of monkeys). They wrote nothing and were eventually chased back out of the region by the Sumerian king Ur-Nammu. Horror always has antecedents. It also has company. The rebels of 2011, for their part, destroyed Shia mosques and Christian graves while also looting Christian churches. What would they have in store for ICARDA? No one knew.

It was the rebels who arrived first at ICARDA. They included men who lived near the facility; they included men who had personally benefited from the center's work. Perhaps as a result of this relationship and the awareness on the part of some of the rebels of the importance of seeds, they decided that if the workers at ICARDA fed them from the center's crops, they would leave the center in peace and allow the workers to continue working. The ICARDA workers had a choice. They could have fled. But they did not. Providing food, they thought, would cost ICARDA little and, at least in the short term, save it much, especially since ICARDA workers do not farm the crops as food—they farm them to study their attributes. They needed only to measure the crops before handing them over to the rebels.[28] The workers at ICARDA were grateful for the peace this deal provided. Amri said of this détente, "We're very lucky that [the rebels] realize the importance of conserving biodiversity; it's one of the activities that has never been interrupted in Aleppo....But we cannot predict how each day will be."[29]

On October 6, 2015, Russian warplanes carried out air strikes near ICARDA's compound. It was reported that the facility was directly hit. On October 7, ICARDA put out a press release saying that the Russian bombs had landed near the facility but not on it. Luck—for now.

· · ·

However one thinks of the future of Syria, its recovery will depend on many steps. Perhaps more war. Definitely diplomacy. But two conclusions are inescapable. One is that if we believe ourselves to be distant from the challenges that agriculture, and food in general, will face in the next decades, the flood of Syrian refugees to countries around the world should remind us that we are not disconnected. Tragedy reconnects us in seconds.

If Syria is to exist again at all as a nation, it will need to reestablish its agriculture. It will need its seeds. This is why, once Ahmed Amri resettled at the ICARDA site in Morocco, he did the unthinkable. He put in a request to the doomsday vault. He requested that the seeds ICARDA had stored in the vault be sent to Morocco and Lebanon so that they could be planted out, increased in number, and then sent around the Middle East, northern Africa, and elsewhere. In an average year, ICARDA distributes twenty-five thousand samples of seeds to those who need them around the world. It plans to continue this work, which would not be possible without the doomsday vault. As Amri said, this continuing mission safeguards "the building blocks for sustaining agricultural development and food security. The CGIAR gene banks are essential for food security; their mission is not controversial, we know how to do it, and it is doable for only $34 million a year."[30] Amri is all too aware of what happens when we collectively fail to achieve food security, as is everyone engaged in the problem of how best to help the hungry people still in Syria as well as the Syrian refugees as they move across Europe.

. . .

The whole Middle East faces prolonged drought. We cannot dither in our attempts to produce new crops that deal well with it. The countries, such as Turkey, with the most water may be fine. Similarly, the affluent countries, such as Saudi Arabia, may simply import food. They may trade, as one article noted, carbon for calories. Saudi Arabia in particular has already decided it has too little water to continue to farm; it will focus on oil and import food. But such a strategy is unavailable to many—perhaps most—of the countries in North Africa and the Middle East. Such countries need to innovate agriculturally now. Here, if we return to the Akkadian empire, we can find one more useful insight into the history of our

condition. While early excavations of the region focused on Tell Leilan, more recent excavations focused on other cities and small towns
in the region. Jason Ur, at Harvard University, contends that these
new excavations show a fate for these small cities that was not as
dramatic as that of Tell Leilan. They may have persisted longer.
Why? They appear, to Ur, to have altered their agriculture prior to
or in response to the drought. Ur sees a shift that included farming
more area and doing so using different techniques (more fertilizer).[31]
In the lesson of these cities, one might find hope (though even this is
debated).[32] Meanwhile, the status quo is that the horrors of the Syrian refugee crisis will be repeated not just within Syria, but among
many of the most populous countries in northern Africa and the
Middle East.

As for Ahmed Amri, he continues to do the work he was trained
to do. There are bad days, worse days, and some good days. Before
the collapse of Syria, he was working on identification of the genetic
resources needed to develop varieties of wheat resistant to the newest wheat pathogen, Ug99.[33] Ug99 threatens wheat in all those countries too poor to use fungicides. It may be that a wheat variety held
at the ICARDA gene bank will save the crops in all those countries.
If it does, millions of people will owe a great debt to Amri and his
colleagues at ICARDA as well as to the Global Crop Diversity Trust
and Svalbard, the Syrians who still work at the site in Aleppo, and
the thousands of others over thousands of years who have tended to
seeds despite all that has happened in the world. Despite all that can
happen. The seeds that, from a tiny grain, can grow enough to feed
a nation, maybe even to feed the future. We have to have hope — for
the Syrians, for those throughout the Middle East, for all of us.
When we look back at the potato famine, we have to wonder: How
it is that no one did more to stop the famine? That no one did more
to understand its causes? That no one did more to prevent it from
ever happening again?

16

Preparing for the Flood

Civilization and anarchy are just seven meals apart.
—Ecologist David Hughes, citing an old Spanish proverb

When I called no one answered. I e-mailed and no one responded. I asked a friend, who gave me the name of another person, who in turn gave me the name of another person. Finally someone answered my e-mail. It was a retired professor. He could meet me, he said, but it might take some time. Or maybe he wouldn't be able to meet at all.

I was asking to see a museum filled with dangerous life forms. It is the museum of plant pathogens, arguably the oldest such collection in the world, in which some of the organisms most deadly to wheat, apples, bananas, and each and every other important plant are stored. The facility contained, I had heard, many thousands of species. I wanted to see this collection, to understand what a gathering of these dangerous forms might look like. It is sometimes noted that there is, somewhere or other, a sample of the smallpox virus in a container at the Centers for Disease Control and Prevention and at a similar facility in Russia.[1] For science. For the future. A Pandora's box, the evil twin of the seed bank in Svalbard.

If we are to save cacao, cassava, potatoes, wheat, and wine grapes from their immediate problems, we need to understand the biology of their worst pests and pathogens. We need that understanding now. In order to get it, we need collections of the pests and pathogens we already know about. We need to be able to compare new problems to those we have already studied. These museums should be grand—ordered and staffed with experts ready when a new problem emerges so that they can identify it, relate its biology to that of organisms we already are familiar with, and begin to take whatever next step is most prudent. In this light, I was excited to see one of the collections on which our future might most depend.

Then I received another e-mail. The man who initially said he would take me to see the collection now couldn't. He had lost the key. No one else seemed to have one. No e-mails for a while. I called some friends of friends who might know whom I could talk to. Nothing. Puzzlement. People who work on plant pathogens at the university where the collection is housed indicated that they had never heard of it. I began to think I was getting the runaround. Then another e-mail came.

The key was found. I received instructions to go to entrance 8B of building 7C. I replied that this sounded great, only to realize when I arrived that these instructions corresponded in no way to a real building. I asked around some more. Someone pointed me in the direction where I found the man. He was holding a key. His wife had just died. He was teetering by the door. His hands shook as he brought the key up to open it. He really couldn't stay long and had come a long way and was retired and so legally should not have been there at all, but he would open the door for me.

I stepped in. Part of me had imagined secured cabinets clearly labeled with the dangers inside. Late blight of potato. DANGER. Witches'-broom. STEP BACK. Maybe there would be alarm bells that would ring when a dangerous specimen had been handled. A bank of computers. Would I have to wear a special suit? Yet as much as I am capable of imagining such a scenario, I knew the likely truth, which is just what I encountered. As I entered the first of what would turn out to be two large rooms, I was surrounded by a hodgepodge of samples. The word *hodgepodge* is generous. In front of me and to each side were exposed plants sitting on the floor and on benches. Some whole tree trunks. Leaves. Stems. Fruit. On each bit of plant material, disease.

When I was writing my book about human hearts, I visited a

collection of diseased body parts in Edinburgh, Scotland. Many of the parts, including hearts, were in jars. Some did not even look like recognizable pieces of a human. Others were inside of the bodies of the humans in which they had been found, broken. This room was something of the equivalent for plants, but rather than just including one kind of disease or one part of a plant, it included everything. Leaves with holes, leaves with spots. Seeds grown into mannish shapes. Stems twisted in contortions. It is only a lack of empathy for plants that prevents this scene from feeling like the evidence locker for civilization's worst crime scenes.

In an old paper written by a trio of great ecologists of the last century, Nelson Hairston, Frederick Smith, and Lawrence Slobodkin famously asked the question, why is the earth green?[2] What, they wondered, prevented pathogens such as those in the museum and herbivores such as butterflies, beetles, mealybugs, and all the rest from eating the green life on earth back down to the ground? Part of the answer would prove to be that plants defend themselves. But perhaps more important, the world is filled not only with species that eat plants—herbivores and pathogens—but also with their enemies, such as the wasps that saved Africa from cassava mealybugs. In the museum room no natural enemies were present, just the pathogens—unadulterated, dangerous, many of them alive, ready to turn the earth back to brown.

The diseased plants were sorted according to whether their diseases were thought to be caused by a pathogen. For much of the history of plant pathology, scholars debated whether sick plants were made that way by the weakness of the plant itself or by a pathogen. It was the old potato blight debate manifest again, this time in a very material way. The collection preserves the old order dictated by this history. Locked in its organization is the history of our understanding and misunderstanding, a history that is still raw and recent. In one of the long rows of shelves in the first room were plants diseased by cancers and mutations—ills intrinsic to them and their environments, things other than pathogens. As with humans, the bodies of plants can just fail, and when they do they have nowhere to run or hide. They are left to face nature's assaults as they wither.

Here was a testimony to such failure. Presumably each such plant also has a human story, that of the person who collected the material or first saw the disease.

In the rest of the rows, though, the majority of plants suffered from extrinsic diseases. Plants killed by fungi came first. Then those killed by bacteria. Then viruses. Somewhere, I suppose, were the nematode worms. I was in a cemetery of plant death organized by the identity of the assailant rather than that of the victim. The collections were not in perfect vials. Instead, like nearly all collections in all of life, they showed the fingerprints of individual human stories. They were in thousands of different vials—or envelopes, depending on who collected the plant. None of it was digitized. The room contained no computer, just a card catalog in which the specimens were identified and classified according to a system that I could not decode and that the old curator, still with me, could not remember. The rules for using the code, or some aspect of the rules, was on a Post-it note that the curator could not find. Then he saw it. It had fallen to the floor. Without it, it seemed, none of the organization made sense.

Many pathogens are fragile in the sense that they depend on the abundance of a particular host. Thousands of species of plant pathogens would go extinct if humans and our crops were to disappear. But unlike a parasite on, say, a hamster, an individual plant pathogen might live for relatively many years, even on a dead branch or in a vial. As a result, even in this room, even without effective preservation, some of the pathogens were alive. Some grew from one sample to another. Some could be seen growing inside drawers. How many and which ones were alive? No one could say. Yet some, if they were given the right conditions, would grow again, springing from plant to plant.

That the specimens are still alive makes them of extraordinary value to science. One can study them to understand what makes some pathogens dangerous and prevents others from being so, which we need to do if we are to predict future dangers. My friends Tom Gilbert and Jean Ristaino used the excellent preservation of plant pathology collections to important ends when they looked at the genes of the late blight from the time of the great famine in Ireland to understand where the late blight came from and why it was so virulent. But because some of these specimens are preserved alive they could reinfect plants. The curator pointed out a piece of wood inside a frame at one edge of the room. That wood, he said, was part of the floor in the old building where the collection was kept. On the wood grew a serpentine monster of a fungus. "That was a fungus from the collection that escaped and started to eat the building," he explained.[3] The same collection, in other words, that could shed light on some of the most significant events in human history could also eat at civilization. The piece of wood had been preserved, the curator noted, because it emphasized the power of fungi and, I supposed, the fallibility of humans.

How could such an important collection exist in such a state? The truth is that the collection I was visiting was not unique. Many collections, be they of plant pathogens or pathogens affecting other organisms, are relatively untended and poorly kept, their newest additions disorganized and stacked here and there where space can be found. Some collections are worse than the one I was visiting. Rotting boxes of unlabeled specimens are not uncommon. It is a near law that the closer geographically a collection is to the regions in which crops and their pathogens and pests are most diverse, the worse shape it will be in. Here I had found very nearly the best-case scenario. And it was an unkempt wilderness of our oblivion.

Collections are unkempt because we have failed to value the men and women able to identify the pathogens affecting plants—or any other species, for that matter. Society doesn't value them much, nor do other scientists who have become, thanks to a reward system that emphasizes flashy discoveries with immediate impact, enamored with the shiny and new and disparaging of the slow, necessary, and long-term. Without great plant pathologists and curators there is no one who really knows the precise value of what is found in a vial, drawer, or envelope. These scholars are dying, and the specimens we use to understand the world aren't well labeled. And, as one scholar pointed out to me, most collections are no longer able to accept new donations, because there's no one around to curate them, to place them where they need to be in the catalog of monsters.

As for the collections of other monsters, the insect pests such as mealybugs, they are even worse. Consider one of the American collections I have worked in recently, for example, a collection that contains perhaps the most complete record of the crop pests and pollinators of its region. This important collection is a record of how our risks have changed through time. One can use its specimens to track how pests or, say, pollinators, responded to changes in climate in the past, or to recent changes in land use.[4] One can use its specimens for many things. Yet, while the collection was staffed by six scientists as recently as ten years ago, only one remains and he is part time. (A new full-time curator will soon be hired. In these stories, someone is always just about to be hired.) And the database for the collection, while sophisticated, is no longer accessible via the webpage designed for the collection (because, well, the database was built by one of those five scientists who left). Such stories are so common as to be the norm, retired curators begging to refill the positions they once held, deans and directors promising to hire someone soon, and the relatively few young curators waiting in temporary jobs until such positions open up, if any in a particular speciality have been trained at all.

Perhaps this is the moment in the story where Bill Gates is supposed to come in, riding on a white horse,[5] to rescue us. He would save the plant pathology and insect collections of the world and fund those scientists capable of identifying, studying, and dealing with them. But Gates did not arrive as I looked around the collection. Instead, the old curator came back. He offered to give me my own key, if he could remember where he had put the extra keys. I said thank you. Then we both went home.

. . .

My people, as one says in the American South, come from the delta of the Mississippi River; their story is one of mud, water, the power of nature, the humility of man. In Greenville, Mississippi, now one of the largest of the Delta towns, the cemetery is full of Dunns. Greenville was founded at an auspicious place on the Mississippi River. Along the Mississippi, as along all big rivers, the places near the banks that sometimes flood are good for farming. Towns are meant to go up high, on hills. But hills are small in Mississippi, the illusion of higher ground in a low realm of cypress swamps, mosquitoes, and floodplains.

In 1835, one of my ancestors, Dr. Thomas Dunn, traveled from Philadelphia to Louisiana and then from Louisiana up the Mississippi River by canoe. Dr. Dunn is said to have been a dentist trained in Philadelphia, though no record of his degree has yet been found. My family explains this as poor record keeping by the dental school. Another interpretation is that he simply never got a degree. He was looking for a place to set up a doctor's office (which is to say that he was setting up shop to work as a family doctor after having maybe, at best, finished a dental degree).

Upon arriving at an auspicious town on the shore of the Mississippi not far from where Greenville now sits, he asked local residents where a good place to settle might be. The people in the houses, upon meeting the man in the canoe—perhaps with a dark sense of humor, perhaps wary of a dentist in doctor's clothing—pointed him farther up the river to the place that would become Greenville after the war. The town grew around the plantation where he settled. He doctored those in need and, because he was a dentist, probably some folks who weren't in need. By the time my grandfather was a boy the city was a burgeoning port. Mark Twain had Huck Finn pause in Greenville; nearly anyone traveling down the Mississippi River would have done so. But the town had a problem. It was too close to the water—water that often rose beyond its banks, into fields, and, well, into Greenville, too.

With time, the Army Corps of Engineers built a levee alongside the river, a bank of dirt to hold back the water. Children played on the levee. They crossed it to throw fishing lines into the swirling brown water. They jigged their cane poles, and sometimes a giant catfish pulled back at the other end of the line. The river could throw up surprises. But when the water got high, the children—the youngest among them, anyway—stayed away. The Mississippi was dangerous on a good day; on a bad day, it would eat anything that came too near. As for the older children and adults, they were stationed along the levee to watch the water. They watched to make sure it did not go over the levee. They watched, too, to make sure that folks from Arkansas, the state on the other side of the river, did not come over and poke a hole in the levee. The only sure way to prevent the towns on the Arkansas side from flooding was to make sure that the rising waters spilled over the Mississippi side.

In the version of the story I remember, my grandfather had the job of watching the river and levee for signs of a flood in the year in which the water rose more than it ever had before, 1927. The year of the great flood. Not *a* great flood, *the* great flood. He did not see anyone from Arkansas, though he looked steadily, shotgun at his side. But he did see the water start to come through the levee. It leaked and seeped, turning the soil before him into a crumbling mush. He had expected a gush of water through a narrow hole into which he might shove his thumb. But the power of the Mississippi is greater than a small boy can imagine; it was greater than the adults of Greenville imagined. My grandfather notified those in charge of what he had seen. Soon others saw other spots where the levee was turning from solid to liquid. Or maybe they saw it first.

In the chaos that followed, stories got tumbled together in the mad heat of nature's wrath. Soon the water crested the levee. People crawled into their attics and onto their roofs. The water poured, then gushed, then came like a crushing monster through town. Hundreds died. Dogs and cows, tied to ropes before the flood, hung dead from trees. Much of the town was destroyed. As for my grandfather, he and his brother got in their boat and went from rooftop to rooftop delivering newspapers and, I suppose, word of what they had seen. The newspaper contained page after page of pictures of those who had died and those who had lost everything. I think about this story often when considering the power of nature.

The story of the flood repeats in many of its details—the rising water, the small figures of people at the levee trying to stop the inevitable. It repeats to such a great extent that it was only in revisiting the history of the Mississippi that I discovered that my grandfather lived through not one but two floods of Greenville, both the great flood and another one, later, in which disaster was mostly averted. What he did and experienced in each flood is muddied in my memory, muddied too among his many retellings, muddied in the waters of the river itself, which has poured over the villages of hunter gatherers and the machines of industrialists alike. We sometimes talk about the end of nature or threats to nature. But nature, though it includes trembling subtleties, can be a son of a bitch. Nature includes the Mississippi River, which will keep flooding, keep testing itself against the efforts of boys, girls, and towns. It also includes the pests and pathogens that appear at the edges of our fields, ravenous and fecund. It includes the riverine forces of ecology and evolution that, if managed imprudently, are entirely capable of doing us in.

. . .

Since the potato famine, plant pathologists and entomologists, armed with collections and knowledge rather than shotguns, have held back the rivers of demons as often as they have failed to. They have done the impossible, again and again. Hans Herren and his colleagues helped save cassava (only to see it begin to falter again), Harry Evans helped save cacao (only to see it begin to falter again), and Jean Ristaino and many of her plant pathology colleagues throughout the world are trying to help to save potatoes as they deal, once again, with unstoppable blight. You don't have to like all these folks; you don't have to want to invite them to dinner to know that we owe them something. By the same token, one has to admit that, protected by these people keeping guard at the levees, our civilization's position is one in which we are behind the levee but below the level of the water. It is a humble position, humbler by the day.

When new challenges emerge—new pests, new pathogens—we need to be prepared. When new climates threaten, we need to be even more prepared. We need this preparedness today, but we will especially need it in a future in which the number of humans is growing. The United Nations projects that the world population will reach 9.7 billion by 2050, two billion more than in 2015.[6] In addition, our dependence on relatively few crops is growing, and the number of new pests and pathogens is growing. Unfortunately, the number of experts trained to deal with pests and pathogens is shrinking just as a generation of great pathologists and entomologists is retiring and dying.[7] No one is filling their jobs. If things continue as they have, the heroes who save us from the monsters gnawing at our corn and blighting our potatoes are going to have to do so from retirement. And then what?

In order to avoid the worst of the flood, we need large teams who know the species around us, be they crop, foe, or friend. The teams need to include people who understand the diversity of crop varieties that might be used to breed new, resistant crops. They need people who understand the archaeology, ecology, and evolution of each crop to figure out where a new pest might have come from. They need to know, too, the forests and grasslands where the wild relatives of our crops live. The team also needs the collections on which their identification and study of new pests and pathogens will be based to be accessible and well curated. We don't want to miss the chance to save ourselves because someone put a key insect specimen in the wrong drawer. Compared to the expenditures we support when crops and countries collapse, the money it would take to do all this is modest—an ounce of prevention compared to the river of impending costs in terms of crops, lives, and civilizations. But we also need something else—a way to connect this team to the farmers who are producing (sometimes) new crops in the field so that they can use new crops when they are developed and (and this is a *very* important "and") can notice when a new problem arises.[8]

Universities such as North Carolina State University, where I work, and Pennsylvania State University, where ecologist David Hughes works, were founded as land-grant institutions. One of the core missions of such institutions, the one most visible to the public, is to educate students. Another is to do research. But the third mission of such institutions, arguably the one with the largest impact on society, is cooperative extension. The cooperative extension service— or simply extension—is the arm of a university charged with engaging farmers. Extension scientists make just the sorts of connections I propose we need. They listen to farmers and take note of their observations from the field. "What is eating your blueberries?" the extension scientist asks. Farmers tell extension scientists about new problems in their fields the moment they arrive. In return the scientists offer farmers the newest available knowledge, the newest approaches to dealing with problems. The information provided by scientists has become increasingly sophisticated, leveraging knowledge of the biology of pathogens and the chemistry of pesticides as well as the best way to combine well-timed pesticide application with a reliance on the enemies of pests and other approaches. Each year tens of thousands of farmers connect to new scientific discoveries through scientists and thousands of scientists learn about new problems from farmers. The relationship improves the well-being of the farmer, their families, and communities and society. It improves the food. It improves the science. It helps hold back the river.

The agricultural extension system has ancient antecedents[9] but was born in its modern form in Ireland in response to the potato famine. Lectures were organized across Ireland at which poor farmers could learn how to save their potatoes (had anyone known, at the time, what to do). The extension system did not save the Irish farmers, but it did make obvious the value of such a system, especially if it could be proactive rather than reactive. In the following decades, the extension system spread around Europe and then to the United States and a few parts of the developing world.[10] But as farms have gotten larger in the United States, the historical goals of the system have become progressively less well served (and, perhaps to some extent as a result, the system has become less well funded). When institutions such as North Carolina State University were founded, half of Americans farmed. Today less than 1 percent do. In a world in which most chickens produced in the United States, for example, are sold and managed by a tiny handful of companies, it no longer makes much sense for extension agents to engage every chicken farm. Small farms are now rare in many of the counties in which extension agents go in search of new observations and insights. Instead, most small farms are now located in developing countries, where farmers are still developing new varieties in the field and where a disproportionate number of new pests and pathogens first appear. We still need extension. We need it more than we ever have, but the nature and geography of the need have changed.

Among the most rapidly growing groups of farmers in the United States is a new generation of individuals growing crops on small plots of land in and around cities (where most people now live and, in the future, nearly all people will live). Many such individuals grow crops for their own consumption or for exchange. Such farmers are finding new ways to connect, both to each other and to the farmers and seed collectors of the past. For example, a large number of farmers and gardeners are now part of the Seed Savers Exchange, a network of individuals dedicated to preserving and actively farming traditional varieties of crops. The exchange maintains more than twenty thousand varieties of crops that are available to its members. These members, some of them growing nothing more than a pot in a window, plant these rare varieties and keep them alive, alive enough to do more than just stay frozen in time. Each time they are grown, the seeds in the Seed Savers Exchange respond to pests and pathogens (at least more so than do seeds frozen in seed banks). They must respond to changes in climate. Theirs is an evolving collection, of the art pieces of antiquity—the seeds—"rehung" in garden after garden, always similar, but never precisely the same. The knowledge members have about the plants they have grown is of great value to universities. Conversely, universities have the potential to help inform the work of the seed savers. But the twelve thousand farmers and gardners involved in the Seed Savers Exchange are not easily served by standard extension systems. They do not fall into standard categories. They are neither horticulturalists nor exclusively crop farmers. They are not farming just berries nor just sweet potatoes. In these ways, the Seed Savers farmers are actually rather similar to another group, a larger group—traditional farmers around the world.

Most small farmers on earth are, and are likely to continue to be in the conceivable future, located in tropical and subtropical countries. It is these small farmers who are most likely to see new pests and pathogens first. The vast majority of species are tropical, and the same is true of those species that eat our crops. It is also the tropical and subtropical farmers who will deal first with the greatest impact of climate change. It is these small farmers who have the most to report back to universities and the rest of civilization about new plagues and pestilence. Ug99, the newest wheat rust fungus, to which most of the wheat growing now around the world is not resistant, first appeared on a small farm in Uganda. Black pod (*Phytophthora* sp.), yet another pathogen that affects cacao, first appeared on a small farm; cassava virus, too, and many more. The farmers in tropical fields are an enormous resource for the future of our collective farms and foods.

What's more, in tropical fields, there is a big need for extension. Globally, farmers lose an average of 40 percent of their crops to pests and pathogens. But on small farms in developing countries, that number can be as high as 80 to 90 percent.[11] Clearly tropical farmers need the insights from universities the most, just as universities need the observations of these farmers the most. But few university extension programs have made the shift from their backyards to the world, and the resources available in the countries with the most small farms are often modest at best. Land-grant institutions tend to pride themselves on having extension agents in every *county* of their home states. We need these agents instead in every *country,* but this shift is very unlikely to happen.

Figure 14. Map of the relative amount of scientific research performed in various countries throughout the world (based on the number of scientific papers produced per capita). The larger the country, the more studies are done in that country. The tropical countries, in which most biodiversity—including crop diversity—dwells, are, relatively speaking, totally unstudied. *Figure by Lauren Nichols, Rob Dunn Lab.*

. . .

To address this need, David Hughes had an idea. Hughes was trained as a basic biologist at prestigious universities. At such universities, those who engage real-world problems are often marginalized or simply not retained, branded with the stigma of utility. Engagement with farmers is far from a priority; it is very nearly a liability, a time-consuming hobby likely to cost one his or her job. As a result, based on his training, Hughes was very unlikely to ever meet extension agents, much less become an advocate for their work. Yet Hughes grew up in an Ireland where the scars of the famine could still be found on the landscape. He saw abandoned potato fields, each one a kind of memorial to the farmers who had died or emigrated. In light of his personal history, in light of the example of Harry Evans, Hughes decided to try to figure out how to reconnect farmers to the newest scientific knowledge and how to reconnect the search for the newest scientific knowledge to the insights of farmers. As the extension system shrinks and plant pathologists and entomologists disappear, the clock is ticking. If he is to do something, he must do so now.

Hughes was inspired by watching his friend Harry Evans in the field, by watching farmers, and then by talking to Marcel Salathé. Salathé is a digital epidemiologist now at the École Polytechnique Fédérale de Lausanne, in Switzerland. Salathé's work often involves engaging participants online, large numbers of participants. He uses social networking data to think about how disease moves through societies. He tracks perceptions of public health and actions related to public health on Twitter. On the surface, he is not the most obvious person for Hughes to collaborate with, but they met and started talking. The talk came easily. Hughes has long studied the parasites of ants. Salathé is fascinated by the ways in which parasites and pathogens shape evolution. Both men, though of very different backgrounds, have spent many hours thinking about the beauty of parasites, how parasites in theory (Salathé) and in the field (Hughes) get their way. Eventually the conversation came around to the future of agriculture and how one might outwit the parasites. They began to develop a plan. They wanted to put, to paraphrase Hughes, Harry Evans in a phone.[12]

For many parasites and pathogens affecting crops, enough is already known to control them, or enough might be figured out with relatively simple studies. We just need a way to get this knowledge out to more people, clever farmers and clever scholars. What if the two men could find someone to build an online portal for information about crops, pests, and pathogens that was accessible anywhere in the world? What if, when people want to know what they are seeing on a plant, they could just look it up? Or even ask or answer a question, with the answers weighted as a function of whether other people found them useful? In an era of *Wikipedia* and Uber, such a portal doesn't seem like a radical idea, yet nothing of the kind existed. Salathé worked with programmers to build the portal, and he and Hughes called it PlantVillage. It was to be the village of the future, where knowledge was shared. But there was more.

Hughes and Salathé would work with other scholars to fill PlantVillage (www.plantvillage.org) with all the information known about each plant pathogen or pest. Again, this seems simple. But the problem was that papers on plant pathogens and pests tend to be available only if you pay for the journals in which they are published, subscribe to them in the way you would subscribe to magazines. For any particular topic this might require subscribing to dozens or even hundreds of journals. More to the point, many of the farmers who most need this information are poor and live in developing countries.

For several decades, scholars have fought to make as much scholarship as possible available to everyone, to open access to it. These fights have been difficult and hard-fought, battles in a longer war. Journals do not want to give readers open access to their information because their business models almost invariably depend on revenue from libraries and, to a lesser extent, individuals who purchase content. Yet much of what can be found in journals is work paid for by tax dollars, paid for by public money in order to benefit the public.

CABI (the Centre for Agriculture and Biosciences International), where Harry Evans works, has begun to put information about plant pathogens online, but the work is all done manually. It must be added to and updated by someone in the central office. As a result, as time passes the data will only be as good as the information that CABI can afford to add to the database. Hughes and Salathé wanted a framework in which, once they got things going, anyone might enter information online about pathogens and pests, information that could be vetted by users. They worked with scientists who helped, in their spare time, to convert the knowledge they found in papers into easy-to-understand information that could be added to the Web portal.[13] The larger the Web portal became, the easier it was to persuade other people to add to its content. It was an extraordinary amount of work—coordinating, persuading, talking, and then, when they were tired, even more of the same. Both men were young assistant professors when they started this project, at a stage in their careers when they were supposed to be doing scientific research and publishing it in obscure journals. They did their PlantVillage work at the potential risk of their careers. They did this because they did not give a damn about who was or was not going to give them credit for doing so.[14] And it worked, or at least the first step worked.

PlantVillage is now the biggest repository of free agricultural knowledge online. It includes information about more than a hundred and fifty crops and nearly two thousand pathogens affecting plants. If you have a smartphone or computer, you can access plantvillage.org. And if you can access PlantVillage, you can access a world of knowledge about the organisms most likely to destroy the plants on which our survival depends. PlantVillage is, in many ways, the direct descendant of the plant pathogen collection I visited. It, too, has rows of pathogens, but rather than being organized by their historical characteristics, the pathogens are organized by host. More to the point, they are organized and curated so one can quickly go to the page for, say, bananas and see information about black leaf streak, anthracnose, Panama disease, bunchy top, banana mosaic, banana aphids, coconut scale, banana weevils, and cigar end rot. Looking up cassava, one finds not only cassava green spider mites but also cassava mosaic disease, cassava bacterial blight, anthracnose, brown leaf spot, white leaf spot, cassava brown streak disease, cassava root rot disease, African root and tuber scale, and other maladies. PlantVillage will never replace insect or pathogen collections. The hope is that it will make those collections more valuable and valued as their utility becomes more obvious to the multitudes. All this is to say that PlantVillage is lovely and useful both to scientists and farmers. Or at least it would be, Hughes and Salathé thought, if people would use it. Would they? Or perhaps more simply, could they?

The vast majority of projects like the one Hughes and Salathé embarked on fail. They fail because it is hard to keep databases up and running. They fail because project leaders get distracted by the other things they have to do (faculty meetings, departmental politics, the need to hire a new department head, the need to appoint an interim department head after the department head that was hired got fired, the need to change the department's name, argue for space, or figure out who is paying for postage when the university cuts funds for mailing things, and so on). They fail because of a lack of funds. But more than anything, they fail because they are not really useful, so no one uses them.

For PlantVillage to work, it needs to do more than provide information. It has to blend the roles of museum collection and extension service: that is, Hughes and Salathé have to have a way to get information to farmers as well as learn from them. To do the latter, Hughes and Salathé set up PlantVillage so that anyone could upload images and insights and questions from their own farms. The two saw this as a way to expand the extension network from thousands (farmers in every county) to billions (farmers in every country). It would allow farmers and those who would like to farm, anywhere in their world, through their everyday interactions, to be sentinels, "recording in real time [any given] disease of crops and animals and tracking it and stopping it before it gets going."[15] As Hughes is fond of saying, each phone now has the computer power of the Apollo missions. More to the point, each new generation of phone has more power than the previous one.

But Hughes and Salathé were even more ambitious. Salathé knew that computer algorithms able to identify people and other features in online images were getting increasingly sophisticated. Facebook can recognize not only you but also your friends. What if Hughes,

Salathé, and their collaborators could gather pictures of pathogens taken by plant pathologists, whether in collections or in the field? Facebook is able to identify you because its algorithm trained itself on the basis of images you have already tagged. If pictures of pathogens come from reputable plant pathologists, Hughes and Salathé would know that the identification of those pathogens is, if not perfect, as good as what one would encounter in a museum collection. They could challenge programmers who develop machine-learning algorithms to create an algorithm capable of correctly identifying as many images as possible. Ideally, an algorithm can not only identify known pathogens but also flag those that seemed new or unusual. This would allow those who have problems with common pests to find the key information online. But when unusual pests and pathogens turn up, those could be sent to experts. As Hughes has said, "The beauty of the phone is that it penetrates society from community gardens in Brooklyn to smallholder farms in Burkina Faso."[16] If all these individuals are using their phones to see, to take "otheries"

Figure 15. Nearly anywhere people live, photos can be taken of plants and used to document (and forestall) the spread of the pests and pathogens that attack crops. The lights of human habitation, as seen by a NASA satellite at night, are a measure of all the places where humans could, in theory, observe where they live and help study the life around us. Today this map is mostly a measure of our impact, but if we act wisely it could one day be a map of our insights. *Data courtesy of Marc Imhoff of NASA/GSFC and Christopher Elvidge of NOAA/NGDC. Image by Craig Mayhew and Robert Simmon, NASA/GSFC.*

rather than selfies, it would be powerful. But whereas the Internet is full of pictures of people, freely available images of plant pathogens are actually rather rare.

. . .

Hughes decided to enlist students at Pennsylvania State University to change that. He had students take pictures of plants of many different species, both those infected and not infected with pathogens. With the help of these students, as well as students at land-grant universities—those with experiment stations where crops can still be found—Hughes was able to gather fifty thousand images of plant pathogens. With these pictures in hand, the most ambitious step in the PlantVillage vision could begin. Hughes and Salathé held a contest in which they offered a prize to the programmers who can develop the algorithm that best identifies the plant pathogens. This is the first major step forward in detecting plant pathogens in a hundred years— that is, if the algorithms work and if anyone uses PlantVillage.

And there is more. Salathé has suggested that in the long run, perhaps just ten or even five years into the future, farmers around the world might, in their phones, have handheld devices able to decode the DNA of the organisms they find growing on their plants. They could, in this way, simultaneously know what they are facing and help immediately update our global understanding of where the challenges to our crops and food dwell.

Of course, the power is greatest when such a network and the information it provides—whether data points, pictures, sequences of genes, or something else—can be connected to people such as Harry Evans and collections of pests and pathogens. When truly new pests and pathogens are found, it takes experts to study and understand them, and there are increasingly few such experts. Sometimes the work of experts needs to be fancy, employing all the latest technologies. In other cases, though, it might be simple, a technology that is known in one place but not another or suggestions based on old studies but never followed up on. When the only focus is on global yields, it is difficult to see smaller local victories. PlantVillage thus has the potential to not only democratize knowledge but also democratize the sorts of solutions we need to look for. In a world in which we can genetically engineer nearly all crops, sophisticated technology can come to seem like the answer to everything. But before you are sure what the answer is, you need to know what the questions are, so that farmers can get both what they ask for and what they need.

PlantVillage will not circumvent the need for better management of our crops; it does not circumvent the need to save seeds and forests and wild cultivars. It starts from a different framework: it starts from the idea that regardless of how well we do these things, we are going to need to find a better way to spot and destroy the enemies of our crops when they come. We need thousands of boys and girls standing at the levee. We need to democratize our ability to see in general, especially to see our foes. It should be possible. Theoretically PlantVillage offers this power, if people use it.

They are. More than three million people have used PlantVillage since its inception — three million people who looked at a plant and saw a problem they could not resolve on their own. Hughes and Salathé expect tens of millions of people to use the site in just a few years. Hughes speaks of billions of users in the future.

He has reason to be hopeful: the algorithms, the ones used to identify pathogens automatically, have begun to show promise. The contest to produce an algorithm to identify plant pathogens produced several that seemed as though they might work. But Hughes and Salathé were impatient, ready for the future, so they also worked with a Lausanne student, Sharada Mohanty, to produce their own algorithm based on approach called deep learning. Their deep-learning algorithm uses data from many pictures to identify the simplest characteristics necessary to tell one pathogen from another. They trained the algorithm using 54,306 photos from fourteen crop species and a total of twenty-six different pathogens that might or might not occur on those species. Having been so trained, a computer using the algorithm could accurately identify the presence or absence of each of those same twenty-six pathogens, when presented with nothing more than a picture of a plant, 99.3 percent of the time. As I said, there is reason for hope.[17]

In the story of agriculture, the levee always breaks. Our only choice is how we are going to be prepared when it does. Nature is not our friend. It is violent, obscene, and remorseless. Nature devours our food, just as the river devoured Greenville. This consumption is its natural tendency. It is only through hard work that it will ever be otherwise, hard work that will require great collections, great plant pathologists, and, if Hughes and Salathé have their way, great algorithms that connect farmers to everything we know through their phones. It will also require something else: farming crops using just the right seeds, the seeds of plants resistant to pests and pathogens, seeds of plants best able to withstand threats. Seeds of great variety, seeds that someone, somehow, must gather, curate, save, and grow in our earth for us all.

Epilogue: What Do I Do?

To be a scientist is to know that most of what is knowable is not yet known. This is the thrill of it all, the great potential for discovery — and it's why I keep going back to work. But to be human is also to be frustrated by our ignorance, especially in those moments in which we need to make decisions. With food, we make decisions every day. We need to figure out ways we can through those decisions, make our food systems more sustainable, more just, and less prone to the total failure of crops, and countries. We need to figure this out even before we understand everything (we will never understand everything).

There *are* some simple things you can do. You can waste less of your food. You can eat less meat. This entire book has been about plants, but eating meat is contingent on the plants being there in the first place to feed to the animals. With some unusual exceptions (such as when chickens are sustained on fodder that would otherwise be waste, including rotting feathers and corn stalks), meat eating is nearly always a waste of food resources relative to eating plants. You can eat local. You can choose foods produced using heritage seeds and produced in agricultural systems managed ecologically. This is particularly important if you live in a region near one of Vavilov's centers of crop diversity. You can also have fewer kids and, if you are in a country with a food surplus, welcome those refugees from countries in which food systems have collapsed.

NEVER OUT OF SEASON

All of these actions can help. But the solutions these life changes favor just reduce the urgency of our race against pests and pathogens. The race does not stop. What else can you do? As I've already noted, if you have a farm, you can help scientists in this race by working with your university extension agent or with PlantVillage to document the pests and, especially, pathogens threatening our crops. But what if you don't have a farm, or even, really, room for a garden? Such is my case. I have a tiny yard with barely enough room left for one more potted plant. In part this is because the yard just isn't very big. In part it is because when writing *The Wild Life of Our Bodies*, I became fascinated with the idea that we could transform cities into places in which fruit trees could help to produce large quantities of food. Somewhat to the chagrin of my neighbors, I ripped up much of my grass and planted peach trees, apple trees, two kinds of fig trees, an olive tree, mulberry trees, a variety of dogwood with edible fruits, a plum tree, cherry trees, and a pomegranate tree. I can start to document the pathogens and pests on these trees for PlantVillage. I will. But when I was talking about all of this in my research group another idea bubbled up, an idea having to do with squash plants, be they pumpkins, acorn squash, bottle gourds, cucumbers, or one of the other plants of the squash family (Cucurbitaceae).

My research group is full of wonderful people in various stages of their careers. Eddy Cruz, for example, is an undergraduate student studying the evolution of the warrens of mammals. Zack Varin, another undergraduate student, is studying the biology of salt. Michelle Musante is studying *Delftia* bacteria that precipitate gold out of solution (really). Megan Thoemmes, a graduate student, studies the species that live in human houses and the nests of chimpanzees. I'm looking for a new student, on a joint project with Ben Reading, to study the biology of caviar. In addition, the group has many postdoctoral researchers. The postdoctoral researchers have all completed their degrees (typically at some other institution) and are in my lab for a few more years of research before they move on to faculty positions or other sorts of jobs. Great ideas can come from anywhere in the research group, and many of the best we have ever had came from the undergraduate students, but this particular idea, the one having to do with squash, bubbled out of the minds of the postdocs.

It is hard to say exactly when the idea arose (it always is), but I can trace a few elements of its origin. It started with dandelions. One of the postdocs in my group, Julia Stevens,[1] was working with teachers and middle school students to try to figure out whether the success of dandelions in extreme environments is possible, in part, because of the unique microbes on their roots (it appears it might be). Julia's work got the whole research group thinking about the beneficial microbes that live on plant roots. Another postdoc, Margarita Lopez-Uribe, meanwhile, was working on squash bees.[2] Much of the squash you have eaten in your life is the fruit of a plant pollinated by a squash bee.[3] Yet we don't even know very much about where squash bees live or why. The species Margarita studies, for example, seems to be absent from Florida, but it may simply be that no one has looked for them there very well. At some point the conversations about squash bees, squash plants, root microbes, and dandelions began to bleed into each other. We began to wonder how much was known about the beneficial root microbes of squash. We began to wonder how much was known about the associates of squash in general.

Then, just as we were talking about how little is known about the associates of squash, Margarita met another postdoctoral researcher, Lori Shapiro. Lori was working at Harvard University and was studying a species of beetle, the striped cucumber beetle, that has evolved the ability to deal with the toxins (cucurbitacins) in the leaves, roots, fruits, and flowers of squash plants and, in doing so, to eat squash. This beetle is a problem in and of itself, but it also carries a pathogen of squash, *Erwinia tracheiphila*.[4] The beetle is found throughout North America and Mexico, but the pathogen is currently found only in the northern United States. If the pathogen spreads it could pose a major threat to squash (including pumpkins, *Cucurbita pepo*—this could be the pathogen that ruins Halloween!). Little though is understood about what limits the spread of this pathogen, or even where it lives, similar to the case of the squash bee. Thanks then to conversations with Julia, Margarita, and Lori, the natural history of squash was very much on my mind. Then came the bottle gourd.

In the spring of 2016, my then six-year-old son appeared to have planted seeds in our backyard in Raleigh, North Carolina. Where the seeds came from, we aren't sure. In any case, some seeds fell, and among the places they fell was between the gravel stones behind our air conditioner, in the one place of dirt we own in which the sun is not blocked out by fruit trees. We were in Denmark for the summer and so none of those seeds were tended to. Yet, amazingly, one of the seeds grew. A single seed. It was a seed of the bottle gourd plant (*Lageneria siceraria*), another member of the squash family. Unlike pumpkins and some other kinds of squash, bottle gourds were domesticated in Africa, yet their biology seems, superficially, similar to that of other squash plants. They grow as vines. They produce large fruits. They evolved to be dispersed (eaten in one place and deposited in feces in another) by megafauna, be they elephants or giant sloths.[5] They grow quickly. When we got back to the house that single gourd plant was thirty feet long and had dozens of flowers. It seemed to threaten to spread over the fence and out into the road (and then it did spread over the fence and out into the road). My son proceeded to watch those flowers every day, to tend to his plant, to note which leaves were being eaten, and to otherwise study it. Having spent time reading Margarita's papers and talking about bees, I naively told him to watch for squash bees. He reported back that he hadn't seen any bees, only moths. This, I thought, must be wrong. But, our collective ignorance of squash in mind, I checked published studies and saw that where they are native, bottle gourds are indeed pollinated by moths.[6] No one appears to have studied bottle gourd pollination in North Carolina. My son's record was a new data point (or it would have been had he taken a picture). It was also the germ of an idea. What if we could get people across the United States (and maybe around the world) to plant key crop species in their yards, but also, in having done so, to begin to help document their associations with pathogens, pests, mutualists, and all the rest?

The good news is that we, in my lab group, knew exactly what to do. We already run a program, Students Discover, through which teachers can work with students to do real science as part of citizen science projects. Such efforts—in which we engage the public, particularly kids, in doing real science — are our bread and butter. We have run projects on belly button biodiversity, the microbiology of homes, ants in schoolyards, and the mammals in backyards. We could, we thought, do a project on squash. In fact, it has already begun. We have begun sending squash seeds across the United States. We send those seeds with guides to planting them, but also with links to online data forms in which observations on pathogens, pests, and mutualists, including pollinators, can be recorded. In some cases, specimens can be sent back to us (for example, if a striped cucumber beetle is found). In other cases, a picture will suffice. We have developed lesson plans that link these planted squash to curriculum goals in the classroom. We have worked with chefs to develop recipes to cook and use what is grown. We have started with one variety of squash, but as the project spreads we will introduce other varieties. (My son, who helped inspire this effort, is planning to plant squash next year, but he will also plant another gourd. As for the gourds he grew this year, they are drying. He wants to make banjos, one for him, others for his friends.)

Our initial goal is to monitor squash plants in every state. You can help us plant squash and, in doing so, keep ahead in the one race that really matters. Our long-term goal, if this all works, is to consider other crops too. Chilies, for example, or cabbage. One of the most exciting things about science is being able to bring people together to answer a mystery in which they have a shared interest. For our squash project, we have all of the experts necessary to help figure out what each of you see on your squash plants and what it means. Margarita Lopez-Uribe, now a faculty member at Pennsylvania State University, will be helping with the study of the squash bees and other pollinators. Julie Urban, also a faculty member at Penn State, will be focusing on the plant-sucking kin of the mealybug that might be found on the plants. Lori Shapiro at Harvard will be keeping an eye on the beetles observed on the plants and the *Erwinia* microbes they carry as well as guiding the overall project. Through PlantVillage, David Hughes and Marcel Salathé will be our partners to consider other pathogens. Angelica Cibrian Jaramillo at Langebio in Mexico will continue the same work in Mexico, where many more species of squash bees (and probably herbivorous beetles) can be found.

Then there is you. You will be the one watching, taking pictures, taking notes, and in some cases collecting samples. You will be part of our attempt to democratize the study of the crops on which all of civilization depends. What we are trying to do is, in our small way, to reinvigorate the land grant system in a model in which the university is connecting not just to farmers, who are now so few, but to anybody with a patch of land. The additional hope, along the way, is to reconnect kids with growing their food, with the biology of food, and in as much as so many kids in the United States still suffer hunger, with food itself.

Watching squash may not seem very radical. After all, our squash project is the logical extension of efforts to make scientific process ever more transparent, accessible, and participatory. Much of the data on which our knowledge of the effects of climate change on birds and plants, for example, are based on projects in which the public actively participates. Yet, engaging the public to help study crop plants, and then making all of the resulting data freely available, stands in stark contrast to the modern agricultural status quo, the status quo of science done by or for agroindustrial companies. The work done in such companies is largely secret. The knowledge such work produces accumulates, but not publicly. Inside each of these companies then is a kind of private knowledge that is growing in size. Just how much of our global knowledge is private is hard to say. More, I'll speculate, than has been the case at any moment in the last hundred years. In this light, the democratization of the study of squash (and then other plants) is indeed something very different. Relative to the dominant trajectory of agriculture, planting and studying seeds yourself is an active step in a different direction. It is a revolutionary step in as much as it will yield new discoveries publicly, but, perhaps even more important, because it has the potential to reconnect our children (and all of us) to the reality that we are connected to our crops and our crops are connected to wild nature, or need to be anyway. Moths came to my son's gourd flowers and pollinated them. It is likely that in some places where children or adults plant squash as part of our project no pollinators will arrive, no squash bees and no honeybees either. When this happens, it will offer us new insight into the biology of squash, but it will also beg the question of where the squash bees might have gone and what you can do to restore them. It will beg the question of how in each small place we can cultivate not just crops, but also the thousands and thousands of other species on which we all depend.

If you plant a squash, you might make a discovery. When and if you do make a discovery, it will be public (and published) that it might be built upon by someone else growing a single squash, or by farmers or scientists anywhere else in the world. By that time, with any luck, you will be sitting down to eat your squash. When you do eat what you grew, think of everything you are connected to, including the farmers and scientists who saved the seeds and have kept the pests at bay, but also the wild species such as the squash bees, and as of yet unstudied root microbes, whose consequences flavor your daily meal. Wherever you live, whatever you eat, you are linked to the rest of life through what you consume, the wilderness that each bite threatens and yet, at the same time, depends upon.

Acknowledgments

Many people helped make this book a reality, either through their inspiration, conversations about the book, or in actually helping to read through sections of the text. I thank a few of those people who I relied on disproportionately here, but the full list of those who have helped to make this book possible would include hundreds. I don't usually suffer writer's block of any kind, but in thinking about how to express gratitude both for the work that is featured in this book (the farming, the research, the desperate attempts to save crops) and by the people who helped me tell those stories, I was stuck. I am more grateful than a few paragraphs will do any real justice, but, nonetheless, here we go.

Thank you to historians Chad Luddington and Matthew Booker for reminding me about relativism and the importance of long, wandering conversations. Matthew read everything; Chad helped especially with the stories of potatoes and of wine. I'm grateful for the help on the wine chapters, even though the whole section on wine, with which Marko Pecaravic and Ivan Pejíc also helped, ended up being cut. Thank you to plant pathologists Jean Ristaino and Marc Cubeta for thoughts about the history of plant pathogens and plant pathology. Thanks to David Hughes, Harry Evans, Jason Delborne, Terry McGlynn, and Colin Khoury for reading and helping with nearly the entire book. Marcel Salethé provided both useful comments and vision. Hans Herren talked to me on the phone and made edits from his vineyard in California. Thank you both for your time and for your work. Peter Neuenschwander provided details about the history of cassava. He also offered a critical eye on the complex story of biological control. Daniel Matile-Ferrero read the chapters on cassava and revisited her notes from decades ago to help reconstruct the early history of the cassava mealybug. Peter Jorgenson and Scott Carroll helped to convene a working group on the evolution of resistance that greatly helped this book, particularly the discussion of resistance and transgenic crops. Thanks to the National Science Foundation for making that working group possible. Yves Carrier, whom I met at that working group, was a huge help in thinking about transgenic crops and the evolution of resistance, as were Bruce Tabashnik, David Mota-Sanchez, and Gabe Zelnik. Dan Bebber helped me to realize that even the very best data on the global pests and pathogens of crops are both incomplete and hard to access.

The story of cacao is a complex one. In addition to the detailed comments from David Hughes and Harry Evans about cacao, the text also benefited from comments or ideas from Jonathan Majer, Jacques Delabie, Wilbert Phillips, Marcellus Caldas, and Rute Fonseca. Stuart McCook provided a historian's perspective on the stories of coffee and cacao. Thank you also to Neil McCoy for helping to tell this and other stories through the production of visuals that make what is hard to imagine hard to deny. Tom Gilbert and Mike Martin told me the stories of their research on potato blight (and Tom's proposed but never funded research on cacao) when they could just as easily have been working on vampire bats or giant squids. Willmer Pérez, David Ellis, and their colleagues at CIP helped me to understand the history of CIP and the amazing diversity of potato varieties that have yet to be well-studied. Gert Kema helped me to understand the story of bananas and their struggles that opens (and provides the cover for) the book.

I went to graduate school with Chris Martine and Mike Gavin who were both students of Greg Anderson. We were all inspired by Greg's work on the domestication and the natural history of crops. Chris helped me to think about the stories of domestication, especially that of potatoes. Mike, whose career has led him to spend thousands of hours thinking about traditional knowledge, read the chapters on rubber and helped me to think about the whole book. Greg read most of the book and provided edits that helped every chapter he touched. Nora Hahn (who doesn't know Mike, Chris, or Greg, but would like them) helped me to think more thoughtfully about the history of agriculture and colonization in Mesoamerica. Joe McCarter and Nigel Maxted provided key insights in the context of the chapter on the wild relatives of crops.

Lee Hickey, Joe-Anne McCoy, Cary Fowler, and Igor Loskutov read, talked about, and then talked about again my treatment of the story of Vavilov, which was also greatly inspired by the work of Gary Nabhan. I look forward to many more conversations, Gary. Thomas Lumpkin at CIMMYT helped provide key insights into the story of his friend Norman Borlaug. Kazuki Tsuji and Tad Fukami provided help with regard to the story of wheat in Japan.

My research group and the colleagues around me every day helped me to think about the natural history of crops and the value of wild nature to agriculture. Steve Frank walks through the neighborhood with me, our neighborhood, often on the way to get beers. Steve noticed mealybugs and scale insects that I overlook and has led me to be more thoughtful about their abundance and consequences. Margarita Lopez-Uribe cued me in to the story of squash bees and, more generally, to the natural history of squash.

Fred Gould deserves special thanks. His ideas run throughout this book, but more than that, Fred created, by force of will and dedication, the Genetic Engineering and Society group at North Carolina State University. This group led the university to hire Jason Delborne, Zach Brown, and Jennifer Kuzma, all of whom helped with this book. It also led me to meet Matthew Booker (for which I am grateful in a general sense, but also as it relates to this book). And it provided me with many hours of colloquia with leaders in genetic engineering, including recent meetings with lead scientists at Monsanto (now Bayer) and Dupont, one of whom gave me hope for the future of agriculture, one of whom didn't.

Harvey Weiss, Ahmed Amri, Pinhas Albert, and Colin Kelley read the chapter on Mesopotamia and helped me to think about the ways in which climate impacts civilizations through its effects on crops. Jason Ur reminded me that the ancient past can contain hopeful lessons (in addition to the tragic ones).

Haris Saslis-Lagoudakis, Romina Gazis, and Vincent Le Guen read and provided comments on the sections on rubber. Nadia Singh and Todd Schlenke provided comments on the Red Queen chapters (including the section about their own work that was later deleted. Sorry...). Cary Fowler offered useful reflections on his own life, but also on the future of agriculture. Bryn Dentiger, Marc Cubeta, Michelle Trautwein, Brian Wiegmann, and Nigel Stork helped me to consider the extent to which the stories of collections that I tell are general ones (they seem to be). Nigel is also the one who first told me the story of the cassava mealybug. Wally Thurman helped me to think about the Borlaug's work through the lens of economics.

One of the things that makes it hard to acknowledge all of those who helped with this book is that I have been thinking about elements of these stories for decades. That thinking has been facilitated by time spent in the Bolivian Amazon with the Dutch group PROMAB and all of its talented scientists (thanks Rene, Marielos, Lourens, and many others). Thank you for giving me a context in which to spend hundreds of hours in tropical agricultural systems and forests—talking about forest regeneration, brazil nuts, cacao, and much more—hours from which this book benefits. It was facilitated by the forbearance of farmers and chiefs alike in Boabeng and Fiema in Ghana (and to the friendship of Audrey Huffman-Reser), my thanks to the people of Boabeng and Fiema. It was helped by the Jelsa Institute for the Study of the History of Life, Civilization, and Wine, which gave me a place to write, work, and think and a context in which to revisit the stories of agriculture in antiquity. Thanks to Núria Roura-Pascual and Xevi, who both gave me a place to write and think and gave me a dinner table to talk about the revolution (Visca Catalunya!). Thanks to the libraries at North Carolina State University and to Susan Nutter and Greg Raschke, who had the vision to make beautiful places in the libraries for faculty to sit, talk, think, and get things done. Thanks to Harry Daniels, my department head, for creating a departmental environment that views engaging the public as a key part of what a scientist does. Thanks to

Susan Marschalk for everything. And thanks to Peter Kjærgaard, Carsten Rabhek, Pernille Hjort, Karsten Vad, and everyone else with whom I work in Denmark for supporting the spread of the work we do to bring science and the public together across Europe.

Thank you to the traditional farmers who engendered the seeds on which our daily bread is based; traditional farmers around the world collectively engendered the most beautiful thing—our collection of seeds—that humans have ever made.

Victoria Pryor helped make this possible. John Parsley lent his keen editorial eye to what needed to stay in the book and what needed to leave. He also helped me to focus in on the most important elements of the story. When I couldn't kill my darlings, he did. Thanks to Michael Noon for his hard work on the entire book, even in the weeks after the birth of his child. Thanks too to the more anonymous editors whose initials I see in the tracked changes at the edges of files that arrive in my inbox, but whose stories I don't know.

And then, as always, the most thanks to my family. To my wife goes the deepest possible thanks, thanks of the sort that leaves me at a loss for words that do any justice. Anything funny in this or any of my other books comes from her edits, as does a great deal of clarity, both in my books and my life. Thank you, Monica. Thanks also for listening to stories, for two years, about farmers, seed collectors, scientists, and all the rest (at one point we had to consciously decide not to talk about the Dark Ages at breakfast anymore). Vavilov never really came to dinner at our house, but it sometimes felt as though he had. And thanks to my children, who give me reason, hope, and joy, and remind me how important it is that we figure out how to sustain the production of our food and to conserve the wild places on which that food depends.

Notes

Chapter 1

1. Even traditional peoples in the least diverse places on earth eat or ate a more diverse diet than most of us do now. The Inuit, for example, are sometimes described as seal eaters because they lived so far north that little else was available. Yet the Inuit also ate walrus, beluga whale, bowhead whale, caribou, polar bear, musk ox, birds, bird eggs, fish, crowberries, cloudberries, grasses, fireweed, roots, and seaweed.

2. This even trickles down to the wildlife in cities. Urban ants and foxes in the United States have bodies in which many of the carbon atoms seem to be derived from the corn and sugarcane in their rubbish-enriched diets. See Clint A. Penick, Amy M. Savage, and Robert R. Dunn, "Stable Isotopes Reveal Links Between Human Food Inputs and Urban Ant Diets," *Proceedings of the Royal Society B: Biological Sciences* 282, no. 1806 (May 7, 2015).

3. Bananas were domesticated in tropical Asia and Papua New Guinea. In these regions, the ancestral bananas and many of the domesticated varieties are pollinated by bats. The plants then produce seeds from which new banana plants grow. But at some point in the domestication of bananas, another approach to growing the plants emerged. They could be replanted from pieces of banana roots, the suckers. This proved much easier than waiting for pollinators and seed production, and, with time, varieties of bananas evolved that were unable to produce seed. They did no worse than those that produced seeds, thanks to humans, and maybe even a little better. They spread. As a result, nearly all the major banana varieties planted outside of tropical Asia are clonal.

4. The Dutch brought coffee to Ceylon as well as to several other islands in the East Indies, including Java.

5. Before the arrival of coffee rust, one-third of the world's coffee came from Asia and Africa. After the rust, less than 5 percent did. See William Gervase Clarence-Smith, "The Coffee Crisis in Asia, Africa, and the Pacific, 1870–1914," in *The Global Coffee Economy in Africa, Asia, and Latin*

America, 1500–1989, ed. William Gervase Clarence-Smith and Steven Topik (Cambridge, UK: Cambridge University Press, 2006), 100–119.

6. Randy C. Ploetz, "Fusarium Wilt of Banana," *Phytopathology* 105, no. 12 (2015): 1512–21.

7. Elected in 1950, Jacobo Arbenz was Guatemala's second democratically elected leader. He proposed to redistribute abandoned banana land to poor farmers in the country and paid United Fruit Company twice what it had paid for the land. Arbenz believed that this would be the first step in creating a better country for his people, a democratic country. The United Fruit Company had other plans. Its leaders persuaded the US government to authorize the CIA to overthrow Arbenz as part of Operation PBSUCCESS. The director of the CIA, Allen Dulles, and the secretary of state, his brother, John Foster Dulles, were both friends of executives of the United Fruit Company. They had even done legal work for the company earlier in their careers. The Dulles brothers helped to convince president Dwight D. Eisenhower of the need to overthrow Arbenz. It was a secret and, from the perspective of the United States, successful coup. As a result, Guatemala's democracy slid into decades of military dictatorship and a brutal civil war. That war would cost the lives of more than two hundred thousand Guatemalans, many of them at the hands of government security forces. The coup also set back democracy in other countries in which the United Fruit Company wielded power. However one attributes blame for these horrors, they, too, are part of the story of the Gros Michel banana.

8. If you want to know what the Gros Michel tastes like, you can go to parts of Asia where Panama disease has not yet arrived and where the Gros Michel can still be grown. Or you can buy artificial banana flavoring. Consumers often say that artificial banana flavoring doesn't really taste like a banana; rather, it tastes more like a banana than a banana does. What it tastes like is the Gros Michel, the prototype used to craft the flavor in the first place.

9. From the Food and Agriculture Organization of the United Nations, at http://faostat.fao.org/.

10. Colin K. Khoury, et al., "Increasing Homogeneity in Global Food Supplies and the Implications for Food Security," *Proceedings of the National Academy of Sciences of the United States of America* 111, no. 11 (2014): 4001–6.

11. One banana expert I talked to, Gert Kema, at the University of Wageningen, suggested that the good news for the banana is that varieties resistant to both to the new variety of Panama disease (*Fusarium*) and to another pathogen, Black Sigatoka (*Mycosphaerella fijiensis*), are being produced using transgenic approaches. The bad news is that such variet-

ies, if they are really resistant, are a decade or more from coming to market. Varieties bred using traditional approaches are even further off. More money, Gert notes, would make all of this go faster, but as he is also quick to point out, more money does not usually materialize until the tragedy is at hand. Meanwhile, we must do all we can to control these pathogens where they have already arrived, and to prevent their spread.

Chapter 2

1. A note here on language: first, a *disease* is the illness that an organism suffers from. For example, AIDS is a disease. Most diseases are caused by biological agents. These organisms are called *pathogens*. The pathogen that causes AIDS is HIV. In the context of the potato famine, late blight was the disease the potato suffered from; whether it was caused by a pathogen would remain to be seen. A *parasite* is, in turn, an organism that lives in or on its host at some cost to that host's fitness. A parasite, such as the larvae of a butterfly on a plant, might weaken its host without causing some specific disease state. The definitional boundaries between pathogens and parasites are fuzzy at best.

2. P. M. Austin Bourke, "The Use of the Potato Crop in Pre-Famine Ireland," *Journal of the Statistical and Social Inquiry Society of Ireland* 21, pt. 6 (1967–68): 72–96.

3. Still, even before the famine the people's healthiness was offset by their near total lack of possessions, including clothes. Children ran barefoot; adults, too. Underwear was nonexistent, and shirts and pants were mosaics of patches and holes.

4. Potatoes lack only vitamins A and D, both of which can be acquired from milk if there is a cow around. Potatoes sustain not only in terms of their quality but also—and especially—in terms of their quantity. Potatoes, sweet potatoes, and other crops that have underground storage organs maintain an advantage over seed crops such as corn and wheat. In theory, one could breed a corn or wheat plant that produced as much food per stem as a potato does, but the plant would fall over. For example, a potato can weigh as much as ten pounds or more. A corncob rarely weighs more than one pound. Picture a ten-pound potato on top of a cornstalk. In this image, you have a precise understanding why root crops are so successful, particularly in places where the amount of food produced per acre must be as high as possible. Like a jellyfish, a potato knows no bones or bounds; it simply grows.
Initially the fecund sustenance of the potato provided the Irish with a way of living better. Potatoes were added to the crop rotation and planted in land that was fallow. By doing this the Irish as much as doubled the

productivity of their land. The potato filled in the fallow spaces; it also filled in the fallow times. Potatoes could be harvested in the summer, when the grains were not yet ready and winter stores had run out. Over the course of generations, however, this luxury turned into dependence.

5. Sometimes rendered just *vastatrix* rather than *devastatrix*. Systematists love to argue about and play with the names of species.

6. Morren was not the first to suggest applying copper sulfate, nor would he be the last. Matthew Maggridge lived near a smelting factory, for instance. He noticed that potatoes growing near the factory's chimneys, out of which billowed copper smoke, tended to be less infected.

7. With C. E. Broom, Berkeley had written the classic "Notices of British Fungi"; he was also the author of "Outlines of British Fungology."

8. In 1879 he donated this enormous collection to the Royal Botanic Gardens, Kew, where it remains.

9. You should pause to contemplate the astonishing recentness of our modern sense of germs.

10. It started not with the potato late blight but with another disease affecting potatoes, leaf curl. Leaf curl arrived in the 1760s in Europe. We now know it to be caused by a virus. In the 1760s people blamed it on the degeneration of potato seeds. After all, the royals of Europe seemed to degenerate over time, their noses ever more crooked and their children ever less capable; why wouldn't the same thing be true of potato seeds? Then in 1841 another disease arrived: dry rot. Again the experts blamed it on the degeneration of the seed, that and bad weather. But Carl Friedrich Philipp von Martius had another idea; he thought that perhaps a fungus was to blame. Martius, the curator of the botanic garden in Munich, was highly regarded. But most other scientists dismissed his hypothesis. His idea that fungus could cause disease—well, it was silly. Martius wrote, "A fungus, fusarium, is the cause of dry rot." Other scientists said no. Martius was right and the other scientists were wrong, but Martius's hypothesis was nearly completely ignored.

11. We now know these organisms to be oomycetes (see page 23) rather than true fungi, a distinction that would not be made for many years.

12. It is perhaps appropriate that Montagne contributed to the confusion over the right name for the species. He is often called Jean Montagne in articles about the history of late blight, but he went by the name Camille. His full name was Jean Pierre François Camille Montagne.

13. In his classic *The Advance of the Fungi*, E. C. Large describes readers of *The Gardeners' Chronicle* as "the prosperous English gentlefolk, who went in for luxury gardens." It was a magazine in which agricultural science of general interest was discussed, but this general interest was

geared toward the affluent rather than those with little more than a potato and a wooden hoe, on whose backs much of European agriculture at the time depended.

14. From John Kelly, *The Graves Are Walking: The Great Famine and the Saga of the Irish People* (New York: Henry Holt, 2012).

15. William Wilde, "The Food of the Irish," *Dublin University Magazine* 43 (1854): 127–46.

16. He had not, but he was trying to figure out whether the weather might have played a role in the severity of the blight and the speed of its spread. He had noted that excessive rainfall might have caused the plants' tissues to be too full of water, but he considered the more likely scenario to be that the blight, having evolved in the wet places where potatoes are native, might spread more readily when conditions are damp.

17. Kingdom Stramenopila, phylum Oomycota, a group typified by its motile, spermlike spores.

18. Here common names get confusing. Blights are sometimes oomycetes, as is late blight of potatoes, and sometimes fungi, as is leaf blight of rubber.

19. E. C. Large writes, vividly, that "if a man could imagine his own plight, with growths of some weird and colourless seaweed issuing from his mouth and nostrils, from roots which were destroying and choking both his digestive system and lungs, he would have a very crude and fabulous, but perhaps instructive idea of the condition of a potato plant when its leaves were mouldy with *Botrytis infestans*."

20. E. C. Large, *The Advance of the Fungi*.

21. A letter in *L'Independance Belge*, August 14, 1845, reprinted in the *Monthly Journal of Agriculture* 1, no. 8 (February 1846): 389.

22. We now know that in the early 1800s, across Europe, farmers used copper sulfate in various mixtures to treat a variety of pathogens. They treated not only seeds and cereals but also, for example, foot rot in sheep. In addition, copper sulfate was used to preserve objects made of wood, including wooden vineyard stakes. It is likely that the value of copper sulfate in dealing with powdery mildew was noticed by many farmers who would have noticed that there was less mildew in areas near their stakes than in areas far from them. See George Fiske Johnson, "The Early History of Copper Fungicides," *Agricultural History* 9, no. 2 (1935): 67–79. Fiske provides an excellent review of the surprisingly fascinating (I swear) history of copper fungicides.

23. Another element of the solution, one adopted more readily though still not until the 1870s, concerned the methods by which potatoes were being grown. Jens Ludwig Jensen, a Danish scientist, used experiments to determine which approaches did and did not forestall the growth of

the blight. Making hills around the potato plants helped keep the sporangia of the blight from washing onto the potato tubers. Getting rid of the foliage before pulling potatoes out so the foliage didn't contaminate the potatoes also helped, as did cleaning out storage bins, particularly those containing infected material. None of this was very sophisticated, yet it was a vast improvement on what was, at the time, being done. It was, in essence, public health and hygiene for potatoes.

Chapter 3

1. John V. Murra, "Andean Societies Before 1532," in *Colonial Latin America*, vol. 1 of the Cambridge History of Latin America, ed. Leslie Bethell (Cambridge, UK: Cambridge University Press, 1984), 61–62.
2. Columbus himself repeatedly bemoaned the fact that although he had clearly found a realm full of many valuable species, he could not tell which was which. As a result, when he did find something he thought he knew, or that he thought might be valuable, he gathered a bunch of it and hoped it could be planted back home. Sometimes he and those who followed him gathered things as a record of the uniqueness of the Americas. On his first voyage, Columbus killed a large snake that he thought the queen would just adore. More often what was gathered were things that could be planted back home, or at least things the conquistadors thought could be planted.
3. Because the conquistadors often married Native American women who then cooked for them, to a great extent Native American women would prove a key factor in what was gathered and what was not. Native American women prepared traditional dishes, but they would have tended to favor those that appealed to the palates of the conquistadors, even though the dishes were composed of native ingredients. Bread made from corn, for instance, to some extent fulfilled the desire of the conquistadors for wheat.
4. A historian whom I asked about this behavior said, "Well, Rob, you know the past is a foreign country," which I take to be his way of saying, "I don't have a clue, either."
5. James Lang, *Conquest and Commerce: Spain and England in the Americas* (New York: Academic Press, 1975), and Geoffrey J. Walker, *Spanish Politics and Imperial Trade, 1700–1789* (Bloomington: Indiana University Press, 1979).
6. The botanic garden established by King Carlos III of Spain in 1788 for testing (acclimatizing) plants before they were transported on to the European continent (Jardín de Aclimatación de la Orotava) still functions in Puerto de la Cruz, on Tenerife.
7. Domingo Ríos, et al., "What Is the Origin of the European Potato? Evidence from Canary Island Landraces," *Crop Science* 47, no. 3 (May 2007):

1271–80; Mercedes Ames and David M. Spooner, "DNA from Herbarium Specimens Settles a Controversy About Origins of the European Potato," *American Journal of Botany* 95, no. 2 (February 2008): 252–57.

8. When planted clonally, via seed potatoes rather than seeds, potatoes don't require pollinators. Yet while this was a useful feature of potato production for the Irish initially, it would prove a problem later, inasmuch as it reduces the diversity possible among potato seeds. Similarly, potatoes are typically not thought to "need" specialist microbes associated with their roots in order to help their roots find nutrients, but it has been shown that potatoes grow many times faster with specialized potato bacteria on their roots than without them. As far as I can discern, no one has studied whether the potatoes introduced to Europe from the Americas brought their specialist beneficial bacteria with them. See J. W. Kloepper, M. N. Schroth, and T. D. Miller, "Effects of Rhizosphere Colonization by Plant Growth-Promoting Rhizobacteria on Potato Plant Development and Yield," *Phytopathology* 70, no. 11 (1980): 1078–82.

9. James Lang, *Notes of a Potato Watcher,* Texas A&M University Agriculture Series 4, ed. C. Allan Jones (College Station: Texas A&M University Press, 2001). See the chapter entitled "The Andean World" for a nice description of the sophistication of Inca agriculture.

10. Similarly, the crops and animals of Europe arrived in the Americas largely without context. Those that survived and prospered were not the tastiest but rather those that were able to complete the entire journey. Columbus, for example, believed that the first colonists he left on Hispaniola failed to thrive (and even to live) because of the absence of European foods. They needed, he thought, fresh meat, almonds, raisins, sugar, honey, wheat, chickpeas, and wine. When he returned, he brought all these things with him, alive. Cows, horses, a few kinds of pigs, and chickpeas all survived. Wheat did poorly and was abandoned. But in each case what he brought was a subset of what he might have brought—a few varieties where thousands were known, varieties Columbus happened to have access to, varieties that are still farmed.

11. The true seeds of potatoes are unpredictable, thanks to sex. Sex mixes up the genes of two parents and helps create offspring that are different from either parent. When conditions are predictable, sex can pose a problem because some sexually reproduced offspring are likely to be less adapted to the environment than would be clones of either parent. But when conditions change (or new pathogens show up), sex is key to producing at least some offspring able to survive the new conditions. In potatoes, sex depends on pollinators, animals that carry pollen from the male parts of one flower to the female parts of another and, in doing so,

fertilize the latter. Perhaps shockingly, we know very little about which species pollinate potatoes in their native range and which pollinate the wild relatives of potatoes. The male parts of potato flowers, like those of many other plants in the family Solonaceae, including tomatoes, only release their pollen when "buzzed" at just the right frequency. It appears that the animals with the right frequency are probably mostly bumble-bees, but which bumblebee species do the best job and whether this has changed through time, along with many other details, remain largely unknown. Most of the relatively few studies on potato pollination have focused on potatoes living in North America and Europe, far from the bees and other insects with which they evolved. See, for example, Suzanne W. T. Batra, "Male-Fertile Potato Flowers Are Selectively Buzz-Pollinated Only by *Bombus terricola* Kirby in Upstate New York," *Journal of the Kansas Entomological Society* 66, no. 2 (1993): 252–54. If you are a student and think it is ridiculous that we don't know more about the pollination of potatoes in the Andes, read the following to get yourself started: C. F. Marfil and R. W. Masuelli, "Reproductive Ecology and Genetic Variability in Natural Populations of the Wild Potato, *Solanum kurtzianum*," *Plant Biology* 16, no. 2 (2014): 485–94.

12. CGIAR's many centers are essential to continued research on tropical and subtropical varieties of crops. The centers also serve as facilities for seed storage, crop breeding, and, as often as not, work that improves conditions for farmers.

13. Lang, *Notes of a Potato Watcher*, 61.

14. For those in the know—blight hipsters, as it were—the particular variety was isolate POX67, belonging to the EC-1 lineage. See also Willmer Pérez, et al., "Wide Phenotypic Diversity for Resistance to *Phytophthora infestans* Found in Potato Landraces from Peru," *Plant Disease* 98, no. 11 (November 2014): 1530–33.

15. Robert Rhoades, "The Incredible Potato," *National Geographic* 161, no. 5 (May 1982): 676.

16. It is said that the annual harvest of potatoes alone, not to mention other Andean crops, exceeds the total value of all the gold and silver brought back from the Andes.

17. Jean B. Ristaino, et al., "PCR Amplification of the Irish Potato Famine Pathogen from Historic Specimens," *Nature* 411 (June 7, 2001), 695–97.

18. Michael D. Martin, et al., "Reconstructing Genome Evolution in Historic Samples of the Irish Potato Famine Pathogen," *Nature Communications* 4, 2172, doi:10.1038/ncomms3172.

19. The results here are a bit complex. What has been found in Mexico and Ecuador is a close, close relative of the late blight that existed in 1845 in

Ireland. The analyses Martin and his colleagues have done suggest that in the fields of Latin America, an even closer relative may still lurk. This relative would be the potato blight's missing link, at least as far as the famine is concerned. It may be extinct; more likely it lives among the tens of thousands of traditional potato fields in Latin America, out there with the great diversity of potatoes and traditional knowledge, all of it largely ignored.

20. Michael D. Martin, et al., "Persistence of the Mitochondrial Lineage Responsible for the Irish Potato Famine in Extant New World *Phytophthora infestans*," *Molecular Biology and Evolution* (2014), doi:10.1093/molbev/msu086.

Chapter 4

1. Pierre Sylvestre, "Aspects agronomiques de la production du manioc à la Ferme d'Etat de Mantsumba (République Populaire du Congo)" (Paris: IRAT [Institut de Recherches Agronomiques Tropicales et des Cultures Vivrières], 1973); Howard Everest Hinton, "Lycaenid Pupae That Mimic Anthropoid Heads," *Journal of Entomology Series A* 49, no. 1 (November 1974): 65–69.

2. Danièle Matile-Ferrero, e-mail message to author, February 2, 2016.

3. The capital of what is now the Republic of the Congo.

4. Danièle Matile-Ferrero, "Une cochenille nouvelle nuisible au Manioc en Afrique équatoriale, Phenacoccus manihoti n. sp. (Homoptera: Coccoidea: Pseudococcidae)," *Annales de la Société entomologique de France* 13 (1977): 145–52.

5. The observation was made by the Commonwealth Institute of Entomology, which suggested it to be a species of the genus *Phenacoccus*.

6. Five different species of mealybug and scale insects were sent, but the abundant one, the new plague, was the species that came to be named *Phenacoccus manihoti*.

7. The word *manioc* derives from the Tupi word for the crop—*mani;* the word *cassava* comes from its Arawak equivalent. Because cassava was farmed throughout the tropical Americas, words for it exist in basically all the languages of the region. The spread of cassava through the region can actually be traced based on the etymology of its various names. See Cecil H. Brown, et al., "The Paleobiolinguistics of Domesticated Manioc (*Manihot esculenta*)," *Ethnobiology Letters* 4 (2013): 61–70.

8. Randolph Barker and Paul Dorosh, *The Changing Agricultural Economy of West and Central Africa: Implications for IITA,* paper prepared for the IITA Program Strategy Planning Review (Ibadan, Nigeria: International Institute for Tropical Agriculture, 1986).

9. Most often, the leaves are brought as a gift of food. But a stem of cassava is also often taken from one place to another as an offering not of food for that day but rather of food that might be planted for the future: "Here, take my cassava, it is sweet."

10. Researchers at a branch of ORSTOM, the Office de la recherche scientifique et technique d'outre-mer, or the French Office of Technical Research Overseas (now the IRD, or the Institut de recherche pour le développement), would go on to document in great detail the biology of the various native parasites and predators in the Congo basin able to eat the mealybug. See Gérard Fabres and Danièle Matile-Ferrero, "Les entomophages inféodés à la cochenille du manioc, *Phenacoccus manihoti* (Hom. Coccoidea Pseudococcidae) en République Populaire du Congo: I. Les composantes de l'entomocoenose et leurs inter-relations," *Annales de la Société Entomologique de France* 16, no. 4 (1980): 509–15.

11. These were the emissaries of the Swiss version of the Green Revolution, coming to spread "the word."

12. Robert van den Bosch, *The Pesticide Conspiracy* (Berkeley: University of California Press, 1989).

13. In some cases, defenses are lost because plants can no longer produce particular toxins. In other cases, the situation is more complex. Wild corn — teosinte — produces airborne chemicals (volatiles) when its leaves are damaged in order to attract predators that eat the herbivores that are in turn eating the maize. North American varieties of corn produce fewer of these volatiles and so are less able to recruit predators when they are needed. See Yolanda H. Chen, Rieta Gols, and Betty Benrey, "Crop Domestication and Its Impact on Naturally Selected Trophic Interactions," *Annual Review of Entomology* 60 (January 2015): 35–58; see also Cesar Rodriguez-Saona, et al., "Tracing the History of Plant Traits Under Domestication in Cranberries: Potential Consequences on Anti-Herbivore Defences," *Journal of Experimental Botany* 62, no. 8 (February 2011): 2633–44; and Amanda M. Dávila-Flores, Thomas J. DeWitt, and Julio S. Bernal, "Facilitated by Nature and Agriculture: Performance of a Specialist Herbivore Improves with Host-Plant Life History Evolution, Domestication, and Breeding," *Oecologia* 173, no. 4 (December 2013): 1425–37.

14. It is even germane to human biology. The developed countries of the world are concentrated in cold regions, where agriculture is more difficult not because of some collective punishment but instead because, it has been contended, in these regions humans are able to escape more of their pathogens. It is hard to imagine any reason anyone would willingly live in Sweden, for example, if it were not for the luxury of not having to deal with malaria and dengue.

15. Petra Dark and Henry Gent, "Pests and Diseases of Prehistoric Crops: A Yield 'Honeymoon' for Early Grain Crops in Europe?," *Oxford Journal of Archaeology* 20, no. 1 (February 2001): 59–78.

16. A related form of enemy release occurs in fields where we spray large quantities of pesticides. These pesticides allow the crops to grow without threat—until, that is, an herbivore evolves the ability to tolerate the pesticide. When this happens, our pesticide-sprayed crops face the same challenges as cassava, but worse. The problem is worse because as long as pesticides are being sprayed, no predators of these new pests can consume the pests, so the pests become the organisms experiencing enemy release. They have all the food they could ever need and nothing in the world to fear.

17. Biogeographers, the folks who study the distribution of species, have tended to ignore the study of species most relevant to humans. As a result, while very good data exist on, say, where a particular species of warbler is located as well as how abundant it is in each place, the data on crop pests and pathogens—even human pathogens, for that matter—are terrible. The best maps available for most pests and pathogens are at the country level and simply record whether a particular pathogen is present or absent. See Michael G. Just, et al., "Global Biogeographic Regions in a Human-Dominated World: The Case of Human Diseases," *Ecosphere* 5, no. 11 (November 2014): 1–21.

18. In addition, Gurr found that whether or not a crop was grown in the British Commonwealth had an effect on the enemies present on it. The most reasonable explanation for this pattern is that after crops were spread by British colonists, the continued movement of crops from one part of the empire to another also tended to move the pests and pathogens affecting those crops.

19. Daniel P. Bebber, Timothy Holmes, and Sarah J. Gurr, "The Global Spread of Crop Pests and Pathogens," *Global Ecology and Biogeography* 23, no. 12 (December 2014): 1398–1407.

20. Donald R. Strong Jr., Earl D. McCoy, and Jorge R. Rey, "Time and the Number of Herbivore Species: The Pests of Sugarcane," *Ecology* 58, no. 1 (January 1977): 167–75; Donald R. Strong Jr., "Rapid Asymptotic Species Accumulation in Phytophagous Insect Communities: The Pests of Cacao," *Science* 185 (September 20, 1974): 1064–66.

21. Strong Jr., et al. "Time and the Number of Herbivore Species."

22. Strong Jr., "Rapid Asymptotic Species Accumulation."

23. Barundeb Bannerjee, "An Analysis of the Effect of Latitude, Age, and Area on the Number of Arthropod Pest Species of Tea," *Journal of Applied Ecology* 18, no. 2 (August 1981): 339–42.

Chapter 5

1. An elegance dwells in knowing that one can pull the pieces of nature apart and predict the result. Ecologists love such elegance, even if it requires abstraction from the muddy details of ordinary life. In this they are like physicists, whose laws all work only in ideal conditions that do not really ever exist. Among ecologists, a favorite example of a trophic cascade is a study in which Tiffany Knight at Washington University in Saint Louis considered eight ponds, four with fish, four without. The ponds were otherwise similar. In ponds with fish, the fish ate dragonfly larvae, which led to fewer adult dragonflies. When the adult dragonflies were rare, they ate fewer bees and other pollinators. With more pollinators, pollination of flowers alongside the pond was more effective. With more effective pollination, the flowers produced more seeds. Does this chain of events happen often in nature? No, or at least not very often, yet it is a simplified illustration of the laws of nature that operate everywhere.

2. By the very most conservative estimates, three out of every four insect species on earth are not named. The situation is worse for fungal species, worse yet for bacteria, and when it comes to viruses, we can't even really guess. As a result, most of the species living in tropical farms, be they cassava, coffee, or chocolate farms, are not yet named, nor is it known whether they are dangerous to those crops, dangerous to humans, or incredibly beneficial. Pest and pathogen species are better known than most species yet ultimately still far from completely understood.

3. In fact another species—another unnamed species, a mite—had begun to kill cassava in East Africa at around the same time.

4. Several years earlier researchers from the biological control unit of the Centre for Agriculture and Biosciences International (CABI) discovered a new mealybug in several countries in the eastern part of the Guyanas. The CABI team included a mealybug expert, D. J. Williams, who thought it was probably the same mealybug that was killing cassava in Africa. But whereas the mealybug on cassava in Africa was pink, this mealybug was yellowish. And whereas the mealybug in Africa reproduced sexually, this mealybug reproduced clonally, via parthenogenesis. The CABI team proceeded to try to release the pests and pathogens that attack this and another mealybug in Africa. But they were unable to get the parasites and pathogens to breed on the cassava mealybug in an insectary in the Congo. This was because, they would later learn, they were working on the wrong mealybug. Herren, who did not yet know any of this complex story, sent specimens to D. J. Williams and Jennifer M. Cox. With Herren's specimens in hand Williams and Cox

realized that what Herren had collected was the same mealybug they had seen in the Guyanas (and that had also been seen in Brazil).

5. Jennifer M. Cox and D. J. Williams, "An Account of Cassava Mealybugs (Hemiptera: Pseudococcidae) with a Description of a New Species," *Bulletin of Entomological Research* 71, no. 2 (June 1981): 247–58. *P. herreni* would soon itself become a problem in Colombia and Venezuela, where its presence is associated with 80 percent crop losses. The story of this species and its emergence has yet to really be understood.

6. It is noteworthy that the mealybugs, while new to the scientists in Paraguay and Bolivia, were not new to the local farmers, who were able to tell the scientists when the mealybugs tend to attack, explain that their abundance varies seasonally, and identify the climatic conditions that tend to make them worse.

7. Hans R. Herren and Peter Neuenschwander, "Biological Control of Cassava Pests in Africa," *Annual Review of Entomology* 36 (January 1991), 257–83.

8. Lopez's wasp was first discovered in Argentina years earlier. What it feeds on in Argentina...well, no one has studied that.

9. If you feel sympathy with the mealybug and worry about its fate, perhaps it is comforting to know that hyperparasitoids of the genus *Prochiloneurus* lay their eggs in the bodies of the wasps inside mealybugs. Nature's justice is layered.

10. G. J. Kerrich, "Further Systematic Studies on Tetracnemine Encyrtidae (Hym., Chalcidoidea) Including a Revision of the Genus *Apoanagyrus* Compere," *Journal of Natural History* 16, no. 3 (1982): 399–430. The species was later transferred to the genus *Anagyrus*. The insects were checked only to see whether they infect honeybees or silkworms.

11. Anna Burns, et al., "Cassava: The Drought, War, and Famine Crop in a Changing World," *Sustainability* 2, no. 11 (November 2010): 3572–3607; Andy Jarvis, et al., "Is Cassava the Answer to African Climate Change Adaptation?," *Tropical Plant Biology* 5, no. 1 (March 2012): 9–29.

12. Interestingly, the diversity is unlikely to get much greater. Crop varieties—such as cassava, potatoes, bananas, and many figs—that are mostly propagated vegetatively (as clones) never get more diverse unless they occasionally get the chance to have sex. In potatoes, it seems that sex occurs—though it has not been well studied—via bumblebees. In their native range, many bananas are pollinated by fruit bats. In their native ranges, most figs are pollinated by wasps. But when these crops move, their chances to have sex decline because their pollinators (fig wasps and fruit bats) are left behind. In addition, in their new homes, the wild plants with which these plants might breed (the crops' wild

relatives) are less likely to be present. In some cases, as with figs and bananas, the ability to have sex at all is lost, the necessary genes broken beyond repair through disuse. In the case of cassava, little is understood about pollination in the wild. Cassava plants growing near their wild relatives clearly exchange genes with those relatives, but even questions as simple as which animals are carrying the pollen back and forth are unresolved. See Kenneth M. Olsen and Barbara A. Schaal, "Insights on the Evolution of a Vegetatively Propagated Crop Species," *Molecular Ecology* 16, no. 14 (2007): 2838–40, and Marianne Elias and Doyle McKey, "The Unmanaged Reproductive Ecology of Domesticated Plants in Traditional Agroecosystems: An Example Involving Cassava and a Call for Data," *Acta Oecologica* 21, no. 3 (2000): 223–30.

13. K. M. Lema and Hans R. Herren, "Release and Establishment in Nigeria of *Epidinocarsis lopezi*, a Parasitoid of the Cassava Mealybug, *Phenacoccus manihoti*," *Entomologia Experimentalis et Applicata* 38, no. 2 (July 1985): 171–75.

14. Much of the hard work of the releases was done by Peter Neuenschwander, Herren's longtime coconspirator in biological control.

15. Herren and Neuenschwander, "Biological Control of Cassava Pests in Africa."

16. Ibid.

17. Although Herren could not have known it, he got lucky in his choice of parasitoid. Or maybe it is better to say that he got lucky in choosing a parasitoid incredibly well suited to the challenge at hand. He did not have the funds necessary to take the parasitoid from field to field across Africa. He would have to depend, to some extent, on the ability of the parasitoid to find cassava plants. But it could go one better than that, as Bellotti and his colleagues would later show. Or maybe it is more accurate to say that together with the cassava plants, it could go one better than that. Cassava plants, when damaged, release signaling compounds. The parasitoid Herren released is attracted to those compounds. In essence, the cassava signals, "I'm hurt by mealybugs," and the parasitoids know where to look. Thanks to this signaling, the parasitoids spread quickly, from damaged plant to damaged plant across Africa, responding to signals in the air—signals few dreamed could exist, much less be able to permeate the dust-drunk plains.

18. Richard B. Norgaard, "The Biological Control of Cassava Mealybug in Africa," *American Journal of Agricultural Economics* 70, no. 2 (May 1988): 366–71.

19. Peter Neuenschwander, "Biological Control of Cassava and Mango Mealybugs in Africa," in *Biological Control in IPM Systems in Africa,*

ed. Peter Neuenschwander, Christian Borgemeister, and Jürgen Lange-wald (Wallingford, UK: CABI Publishing, 2003), 45–59.

20. Ker Than, "Parasitic Wasp Swarm Unleashed to Fight Pests," *National Geographic*, July 19, 2010, at http://news.nationalgeographic.com/news/2010/07/100719-parasites-wasps-bugs-cassava-thailand-science-environment/.

21. An article about Matile-Ferrero's (still active) career notes that no one has been hired to replace her now that she has retired. Nor, the article notes, is anyone in charge of the group of insects Matile-Ferrero works on (which includes not only mealybugs but also scale insects, aphids, and whiteflies) at the major collections in Paris, London, the United States National Entomological Collection (in Maryland), and the Bohart Museum of Entomology at the University of California, Davis. Matile-Ferrero herself suggests that there is a good network of younger people trained in her field (coccidology), but whether they are able to find long-term work is probably an open question.

Chapter 6

1. CEPLAC was founded in 1957 as a "temporary commission."
2. Marcellus M. Caldas and Stephen Perz, "Agro-Terrorism? The Causes and Consequences of the Appearance of Witch's Broom Disease in Cocoa Plantations of Southern Bahia, Brazil," *Geoforum* 47 (June 2013): 147–57.
3. It also helped usher in a style of cacao farming that, although more productive in terms of the quantity of cacao produced per acre (nearly double), was far more negative than positive in its effects on the environment. In addition, this style of farming made the cacao more susceptible to pests and most pathogens than it had been before.
4. Though, as it turns out, things are not quite so simple.
5. Harry Evans reports that the town of Vinces, Ecuador, even has a small, derelict Eiffel Tower. Here was, the message seemed to be, an Ecuadorian Paris.
6. Elizabeth Keithan, "Cacao Industry of Brazil," *Economic Geography* 15, no. 2 (April 1939): 195–204.
7. Before humans arrived in the Americas, cacao and its relatives were dispersed by monkeys, as is still the case for wild cacao pods not collected by humans. Giant sloths and gomphotheres (extinct kin to the elephant) may also have dispersed the seeds along with some of the pathogens. See Paulo R. Guimarães Jr., Mauro Galetti, and Pedro Jordano, "Seed Dispersal Anachronisms: Rethinking the Fruits Extinct Megafauna Ate," *PLoS One* 3, no. 3 (March 2008): e1745.

8. In the early years of cacao production in Bahia, many of those who grew Amelonado cacao were small farmers, including many who immigrated to the region. But with time, and given the high demand, cacao production became an endeavor carried out by a few relatively affluent "colonels." In the succeeding decades and centuries, demand for cacao on the global market waxed and waned, but inevitably with less consequence for the colonels than for the owners of small farms or for the poor people who worked for the colonels. When the price dropped, the owners tended to fire their employees and leave cacao unpicked, in pods, growing like tumors from their trees' trunks. When the prices rose, they hired and again and had every last pod, and every last bean within them, picked. These boom-and-bust cycles reinforced and exacerbated income inequalities in the region. See Keith Alger and Marcellus Caldas, "The Declining Cocoa Economy and the Atlantic Forest of Southern Bahia, Brazil: Conservation Attitudes of Cocoa Planters," *Environmentalist* 14, no. 2 (1994): 107–19, and Caldas and Perz, "Agro-terrorism?".

9. Fungi and plants are far more closely related than bacteria and plants or viruses and plants, so many things that kill a fungus harm the host as well.

10. H. Laker and S. A. Rudgard, "A Review of the Research on Chemical Control of Witches' Broom Disease of Cocoa," *Cocoa Growers' Bulletin* 42 (1989): 12–24.

11. L. H. Purdy and R. A. Schmidt, "Status of Cacao Witches' Broom: Biology, Epidemiology, and Management," *Annual Review of Phytopathology* 34, no. 1 (February 1996): 573–94.

12. Fortunately, an international group had already been charged with helping develop approaches to control witches'-broom. Unfortunately, that group had noted, "If this disease were ever to hit Bahia's cacao area, [the production of cacao] would decline severely. This...might well cause the global market to explode." See *Managing Witches' Broom Disease of Cocoa* (Brussels: International Office of Cocoa and Chocolate, 1984).

13. Traditionally, Amazonians consumed the fruit of the cacao pod rather than the seeds. It was only once cacao was domesticated and moved to Mesoamerica that the seeds began to be consumed. Most of the size and weight of the cacao seed is attributable to the two small leaves inside the seed that will grow once it germinates, the cotyledons. These cotyledons are full of fat, which they will use during germination. Fat is the best way to store a lot of energy in a small space, which is what the seeds need to do if they are going to grow on the forest floor, where sunlight is scarce. It is this fat, along with all the nutrients stored up in the leaves for defense and other roles, that gives the chocolate its lovely flavors. Other flavors come from the fermentation of the beans that occurs on the ground where

they are grown. The beans yield all the products you can find in a chocolate bar. Cocoa butter is produced by squeezing the seed of the chocolate
so that the fat drips out. Cocoa powder is what you get when you grind
up the beans that have had the cocoa butter squished out of them. Chocolate (also called cocoa liquor) is produced by roasting the beans and
grinding them up.

14. Nor is it yet known how witches'-broom causes disease. We know that the
lineage that includes witches'-broom is composed primarily of species that
are able to break down only dead plant materials. It also appears that the
species of fungus associated with witches'-broom has acquired genes that
allow it to better degrade living tissue (it stole them from a species of *Phytophthora* oomycete). But some argue that the fungus thought to cause the
disease is, even with these genes, not sufficient to trigger all the symptoms
of the disease when inoculated into plants. Instead, witches'-broom may
be caused by the infection of the fungi along with another organism that
rides with the fungi into the plant—a virus, perhaps.

15. See J. L. Pereira, L. C. C. De Almeida, and S. M. Santos, "Witches'
Broom Disease of Cocoa in Bahia: Attempts at Eradication and Containment," *Crop Protection* 15, no. 8 (December 1996): 743–52.

16. The knot is described in the documentary *The Knot: Deliberate Human
Act* (*O nó: Ata humano deliberado*), directed by Dilson Araújo: https://
www.youtube.com/watch?v=_omPiYocm-4#t=716&hd=1.

17. The cacao plantations of Bahia were within the region that was once
the Atlantic Forest of Brazil. When the first Europeans landed in Brazil,
this covered an area no smaller than half a million square miles, an area
twice the size of France filled with species found nowhere else on earth.
Today, as a result of agriculture, logging, human population expansion,
and other pressures, just fifteen hundred square miles remain. The loss
of nearly all this once great forest has been called one of the greatest
biological tragedies of our time. The remaining patches of trees are
home to tens of thousands (and perhaps hundreds of thousands) of bird,
mammal, plant, and insect species that are found nowhere else. The
traditional plantations in Bahia provided corridors among patches of
forest through which animals could move as well as forests and habitats
in their own right—thousands of square miles of forest canopy in a
region where every patch matters. As forests, traditional cacao plantations helped conserve tens, perhaps hundreds, of thousands of species.

18. Where the definition of agricultural terrorism is construed broadly as
"the intentional use, by any human agent other than uniformed military personnel, of organisms (or their products) to cause harm (or
death) to humans, animals, or plants." See Laurence V. Madden and

Mark L. Wheelis, "The Threat of Plant Pathogens as Weapons Against US Crops," *Annual Review of Phytopathology* 41, no. 1 (September 2003): 155–76.

19. Neil M. Ferguson, Christi A. Donnelly, and Roy M. Anderson, "Transmission Intensity and Impact of Control Policies on the Foot and Mouth Epidemic in Great Britain," *Nature* 413 (October 2001), 542–48.

20. Described in the documentary *The Knot*.

21. Ibid.

22. Perhaps the only thing, some Brazilians began to suggest, that would save Brazilian cacao was if witches'-broom arrived in West Africa, making all the world's cacao sick and, however horribly, evening the playing field. It was only a matter of time, some said, almost as a threat.

23. By his own account, though he has little reason to lie. He is a free, unindicted man, confessing guilt and, in so doing, putting himself at potential risk both legally and from anyone who suffered the fall of cacao and wanted to exact retribution.

24. As early as 1991, long before Timóteo came forward, some suspected the involvement of the Workers' Party in the spread of witches'-broom, but at the time the accusation was little more than speculation.

25. The forest on the border with Bolivia had been cut under the orders of former president General Emílio Médici, who wanted the Amazon cleared for agriculture. He also had the Trans-Amazonian Highway constructed, along which cacao was to be a key crop. The aim for cacao was that its production would more than double in Brazil and that more than half of all production would come from the Amazon plantations. Yet the supposedly resistant varieties of cacao (resistant, at least, when grown in Trinidad) that were planted quickly became overcome with witches'-broom, and the plantations suffered losses of up to 90 percent. Had that forest not been cut, had those misguided plantations not been planted, it would have been much harder to move the witches'-broom.

26. Article 109, item IV, of the criminal code.

Chapter 7

1. As much as 70 percent of all cacao in the world comes from either Ivory Coast or Ghana, but cacao is grown in more than fifty countries, where it provides income for more than forty million people.

2. Adrian Frank Posnette was his birth name, but he was known as Peter by all those who worked with him.

3. The new imports could have brought witches'-broom with them, but fortunately none did. Care had been taken to quarantine the imports at the Royal Botanic Gardens, Kew, and then at the University of Reading,

an effort similar to the quarantine for the parasites and pathogens affecting the cassava mealybug.

4. Girolamo Benzoni, an Italian traveling on the ships of Spanish conquistadors, wrote in 1556 that "cocoa flourishes only in a hot climate, in shaded locations; if it were exposed to the sun it would die." Benzoni saw cacao being planted in low densities beneath rain-forest trees. The cacao trees in these traditional systems yield "shade chocolate," much as coffee plants farmed in similar conditions yield "shade coffee."

5. Diane W. Davidson, et al., "Explaining the Abundance of Ants in Lowland Tropical Rainforest Canopies," *Science* 300, no. 5621 (2003): 969–72.

6. These entomologists included Dennis Leston, from Imperial College; Jonathan Majer, who would go on to study the ants of Australia; and Barry Bolton, who produced *the* catalog used by entomologists to look up the names of the world's ants.

7. This led to a series of papers, the first of which was "Thermophilous Fungi of Coal Spoil Tips: I. Taxonomy," *Transactions of the British Mycological Society* 57, no. 2 (October 1971): 241–54. The second was "Thermophilous Fungi of Coal Spoil Tips: II. Occurrence, Distribution, and Temperature Relationships," *Transactions of the British Mycological Society* 57, no. 2 (October 1971): 255–66.

8. Harold Charles Evans and Dennis Leston, "A Ponerine Ant (Hym., Formicidae) Associated with Homoptera on Cocoa in Ghana," *Bulletin of Entomological Research* 61, no. 2 (November 1971): 357–62.

9. On the genera *Camponotus* and *Crematogaster,* see Harold Charles Evans, "Transmission of *Phytophthora* Pod Rot of Cocoa by Invertebrates," *Nature* 232 (July 1971): 346–47.

10. There now appear to be several species of *Phytophthora* black pod oomycetes infecting cacao, two in different regions of West Africa and another in the Americas. It is likely that more species will be found in the future, both as the pathogen is better studied and as new varieties and species make the jump from native forest plants onto cacao. Viruses that infect (and attack) these oomycetes are now being sought, but even once they are found, other viruses are likely to be needed to control the various species of *Phytophthora*. See Harold Charles Evans, "Cacao Diseases—The Trilogy Revisited," *Phytopathology* 97, no. 12 (December 2007): 1640–43; and Harold Charles Evans, "Invertebrate Vectors of *Phytophthora palmivora,* Causing Black Pod Disease of Cocoa in Ghana," *Annals of Applied Biology* 75, no. 3 (December 1973): 331–45.

11. Because these fungi evolved to manipulate the immune systems of insects (in order to prevent themselves from being attacked), they have

proved useful in a variety of medical contexts. The drug cyclosporine, for example, used to suppress human immune systems for organ transplants, comes from one of these fungi.

12. The fungi invade the living bodies of ants and fill them with fungal cells. Once established in the ants, the fungi stimulate their hosts to leave their nest. The fungi then wait for the ants to arrive at a place that, from the fungus's perspective, is just right—not too dry, not too wet, not too hot, not too cold. Once there, the fungi destroy the muscles that hold the mandibles of the ants open. In most species of ants, holding mandibles open takes energy, whereas snapping them shut does not. As a result, once their muscles are destroyed, the ants bite down into whatever they are standing on, which if things work out well for the fungi is a leaf of some sort. There the fungus grows a stalk up into the air and releases spores. Scientists interested in the basic functioning of the world have studied these fungi for a century to understand their strange ways. To these scientists, including Hughes, the fungi are beautiful manifestations of the elaborateness of nature—its wickedness, too. They are just one example of the many ways parasites can control the behavior of their hosts in order to benefit themselves. To cite another example, hairworms cause crickets and other insects to dive (fatally) into water. The hairworm needs to get into water to mate. The parasite *Toxoplasma* causes rats and mice to be attracted to the smell of cat urine, thereby increasing the odds that they are eaten. (It also, we now know, has a similar effect on chimpanzees, causing them to be attracted to the smell of leopard urine.) *Toxoplasma* can only successfully have sex inside a cat.

13. Around every corner in the story of West African cacao are more mysteries. The many varieties of cacao swollen shoot virus suggest that the virus colonized cacao from multiple hosts, though just which hosts, and how frequently, is not yet clear. The varieties of mealybugs that transmit these viruses have changed through time; the mealybug that used to transmit the virus is now rare. Just why is unclear. Climate change has been suggested, though not studied, as a possible cause. No fewer than a dozen mealybug species, and likely several unnamed mealybug species, now transmit the virus.

14. Dennis Leston, "The Ant Mosaic—Tropical Tree Crops and the Limiting of Pests and Diseases," *Pest Articles & News Summaries* 19, no. 3 (1973): 311–41.

15. Similarly, the broad-scale fungicides that kill oomycetes are also likely to kill beneficial root fungi and leaf endophytes, particularly when the fungicides are sprayed from the ground up, toward the tops of tall cacao trees.

16. These practices are often described as backward or simple in comparison to larger, more intensive cacao plantations. Or at least that is how they are described, again and again, until pests and pathogens arrive.

Chapter 8

1. After first attending the Moscow Commercial Institute, per his father's wishes, and graduating in 1906.
2. It was to be published Russian just five years later, in 1864.
3. Quoted in Igor G. Loskutov, *Vavilov and His Institute: A History of the World Collection of Plant Genetic Resources in Russia* (Rome, Italy: International Plant Genetic Resources Institute, 1999).
4. The seminar, entitled "Genetics and Its Linkage with Agronomy," was one of several he gave as part of the Golytsin Higher Agricultural Courses for Women. In it he discussed genetics, the value of pure (inbred) lines in breeding, mutations, and the potential for using the theory of genetics in the practice of crop breeding. See Loskutov, *Vavilov and His Institute*.
5. Vavilov would go on to show that it was actually a new species of wheat, which he named *Triticum persicum*.
6. Throughout his decades of travel, Vavilov would never be without his fedora. No matter if he was hiking in mountains, deserts, or jungles. No matter if he was fleeing crocodiles or bandits. No matter if he was trapped in a tent with snakes. Vavilov would not be seen without his suit, a well-tied tie, and his perfectly placed hat.
7. Vavilov had seen similar symptoms in farmers after they ate bread in his earlier travels on Persia's northern border and knew their cause. See Peter Pringle, *The Murder of Nikolai Vavilov: The Story of Stalin's Persecution of One of the Great Scientists of the Twentieth Century* (New York: Simon and Schuster, 2008).
8. Such similarity was not a matter of chance; it had evolved in response to the advantages conferred on any crop able to persuade a farmer to unwittingly plant it along with his chosen seeds. Vavilov would be the first to study this kind of mimicry, which has since proved to be very common. Any weed that can mimic a crop seed has the potential to get planted by farmers in field after field: natural selection strongly favors very good mimics.

    The first step in such mimicry is often crude—the evolution of a larger seed size (which makes seeds more likely to end up on the right side of the sieve as they are sorted). But such mimicry can also be extreme. The plant false flax (*Camelina sativa*) has even evolved into two different varieties, one a mimic of the oilseed variety of flax, another of the variety of flax used in making textiles. In some cases, these mimics become, over time, domesticated themselves. The oilseed version of false flax, for example, is a minor crop. More significantly, rye appears to have originally been a mimic of wheat. It grew among its stems and was only later (and perhaps begrudgingly) domesticated, an insight Vavilov formed on his 1916 expedition. He developed this hypothesis based on his observation that as he went up in elevation during his travels, fields of winter wheat were increasingly contaminated with ancient-seeming varieties of rye.

Often mimics become agricultural plants that grow in extreme conditions, as is the case with sand oat, which grows on sandier soil than does any true oat.

9. Perhaps he used a technique that has effectively led to the escape of many field biologists from border guards around the world: talking at length about the plants he collected and, in doing so, boring the detector into exhaustion.

10. See more about the story of this wheat variety in Gary Paul Nabhan, *Where Our Food Comes From: Retracing Nikolay Vavilov's Quest to End Famine* (Washington, DC: Island Press, 2008).

11. Translation from Loskutov, *Vavilov and His Institute*.

12. By then also beginning to be known as the Department of Applied Botany and Plant Breeding.

13. Loskutov, *Vavilov and His Institute*.

14. Especially the work of the great Swiss biogeographer Alphonse de Candolle, e.g., *Origin of Cultivated Plants* (New York: D. Appleton, 1885).

15. In Nikolai Ivanovich Vavilov, *Origin and Geography of Cultivated Plants*, trans. Doris Love (Cambridge, UK: Cambridge University Press, 1992).

16. For a better sense of Vavilov's journeys, Gary Nabhan's book *Where Our Food Comes From* is a must-read. Nabhan visits many of the places Vavilov went and, in doing so, considers how traditional knowledge and the use of crop varieties have changed since Vavilov's time. Much, Nabhan concludes, has been lost in many regions, but much also remains, the persistent flowering of cultures and their seeds.

17. The Bureau of Applied Botany already had a long history of seed collection, crop breeding, and agricultural research in general. The focus of the bureau, however, was on the Russian Empire, which given Russia's geographic dimensions was nonetheless an ambitious mandate. By 1914, the collection already included some fourteen thousand seed accessions (collections) from across Russia and beyond. For an excellent review of this history, see Loskutov, *Vavilov and His Institute*.

18. He said this before dismissing the reporter with "Apologies, monsieur, life is short, time has no patience" (Pringle, *The Murder of Nikolai Vavilov*, 188).

19. Estimate from Nabhan, *Where Our Food Comes From*.

Chapter 9

1. Carl-Gustav Thornstrom and Uwe Hossfeld, "Instant Appropriation—Heinz Brücher and the SS Botanical Collecting Commando to Russia 1943," *Plant Genetic Resources Newsletter* 129 (2002): 39.

2. He wrote, "The climatic conditions of these eastern areas place very special demands on cultivated plants....Falling back on the primitive origins of the cultivated plants that Vavilov collected is all the more important in the current state of plant genetics." As the writer Noel Kingsbury points out, the Nazis were also interested in new plant varieties. These included homeopathic herbs and a potential substitute for rubber, the dandelion *Taraxacum kok-saghyz*. These new crops were often grown in nurseries in concentration camps. See Noel Kingsbury, *Hybrid: The History and Science of Plant Breeding* (Chicago: University of Chicago Press, 2009).

3. One of many tragic ironies was that Vavilov had been arrested not so many years before, in Mexico, and accused of the opposite—stealing seeds for the Russian government to the detriment of other countries, including Mexico and the United States. See Gary Paul Nabhan, *Where Our Food Comes From: Retracing Nikolay Vavilov's Quest to End Famine* (Washington, DC: Island Press, 2008).

4. Gary Paul Nabhan puts it plainly enough when he says, "Lysenkoism dragged Soviet biologists and agronomists into the murky backwaters of the life sciences until Lysenko's shoddy experiments and ideological rants were discredited by an overwhelming outcry from other Soviet scientists in 1964." See Nabhan, *Where Our Food Comes From*.

5. Mark Popovsky, *The Vavilov Affair* (Hamden, CT: Archon Books, 1984).

6. Meanwhile, the two friends whom Vavilov had falsely accused of being "agricultural spies"—one of whom Vavilov had encouraged to return to Russia, despite the man's being happily settled in the United States—were executed.

7. Kasha (*Fagopyrum esculentum*), Vavilov would have noted, is also called buckwheat. It is not, however, a wheat at all, or even a grass, but rather a kind of sorrel. It was domesticated somewhere in Asia and during Vavilov's life became an important Russian crop.

8. Kingsbury, *Hybrid*.

9. Nabhan, *Where Our Food Comes From*.

10. Calvin O. Qualset, "Jack R. Harlan (1917–1988): Plant Explorer, Archaeobotanist, Geneticist, and Plant Breeder," in *The Origins of Agriculture and Crop Domestication: The Harlan Symposium*, ed. A. B. Damania, et al. (Aleppo, Syria: 1997).

NOTES

Chapter 10

1. Minnesota was at the time probably the world leader in the study of wheat. John Perkins argues that this preeminence is in part attributable to the role of the state in the milling and baking of wheat growing on the Great Plains. See John H. Perkins, *Geopolitics and the Green Revolution: Wheat, Genes, and the Cold War* (Oxford, UK: Oxford University Press, 1997).

2. In 1953, a strain of stem rust (15B) would destroy 80 percent of the durum wheat growing in the United States.

3. Stakman was called to Mexico by Henry Wallace, Franklin Delano Roosevelt's vice president. Wallace had recently visited Mexico and had been struck by the need for the improvement of crops, particularly corn and wheat. Wallace had several reasons to hope for such improvement. For one, he really appears to have believed it was the right thing to do, a just and important cause. For another, stabilizing the agriculture of Mexico was seen to be vital in the fight against Communism (poverty and instability, it was said, bred Communism). Finally, Wallace was the president of Pioneer Hi-Bred, the company that had brought new hybrid corn varieties to North America. The company could serve as a model for the next steps in Mexico and maybe even derive some benefit for itself. For all these reasons, Wallace persuaded the Rockefeller Foundation to bring a team to Mexico; it was the Rockefeller Foundation that contacted Stakman.

4. Those who worked with Borlaug in Mexico note that he expected the same of everyone who worked with him. To work with Borlaug was like joining the Peace Corps and the Marines at the same time, said Jesse Dubin in an interview with Susan Dworkin. The only difference was that the entire focus was on wheat, and the enemy was either rust or, more often, inefficiency—inefficiency in work and inefficiency in how much food the wheat could produce from sunlight, irrigation, and fertilizer. At night the men relaxed by drinking and talking about wheat. Then they woke up and worked on wheat. See Susan Dworkin, *The Viking in the Wheat Field: A Scientist's Struggle to Preserve the World's Harvest* (New York: Walker Books, 2009), 239.

5. This was to be the precursor to the International Maize and Wheat Improvement Center, CIMMYT (or Centro Internacional de Mejoramiento de Maíz y Trigo). Agricultural history is full of acronyms. CIMMYT would be the center through which US scientists worked with Mexican scientists to breed new varieties of wheat and corn. In time, it would become part of a consortium of centers, each with its own acronym, carrying out similar missions in other regions of the world, including the International Rice Research Institute, in the Philippines

288

NOTES

(IRRI); the International Potato Center, in Peru, which I have already mentioned (CIP, or Centro Internacional de la Papa); the International Center for Tropical Agriculture, in Colombia (CIAT, or Centro Internacional de Agricultura Tropical), where Tony Bellotti worked when tracking down the cassava mealybug; the International Institute of Tropical Agriculture (IITA), where Hans Herren worked; and the International Center for Agricultural Research in the Dry Areas (ICARDA), to which we will turn later. The group to which these institutes all belong is the Consultative Group for International Agricultural Research (CGIAR), originally funded by the World Bank, the Food and Agriculture Organization (FAO), and the United Nations Development Programme (UNDP). Then, of course, there is also CABI, the Centre for Agriculture and Biosciences International, which has played a major role in many aspects of crop protection. I've tried to avoid using these acronyms where I have been able to, but if you're ever at a party with the small group of people in the know, you will need to know them as well. Consider this endnote, then, to be your cheat sheet. And if you really want to seem as though you are in with the in crowd, don't say "CGIAR": use the short form of this acronym, "the CG system," which frankly makes it sound like the sort of weight-loss product you might buy late at night after watching an infomercial.

6. The wheat varieties from Mexico that Borlaug used were developed by Mexican farmers from those brought over by Cortés and other Europeans in little more than five hundred years. They are a testament to the extraordinary ability of traditional farmers to create diversity from simplicity. In many ways, this was the opposite of what Borlaug was doing, with one exception: Borlaug saved not only the traditional varieties he collected but also the new varieties he made through his crosses. Together, the traditional and new varieties numbered around five thousand. Borlaug sent the seeds to the National Small Grains Collection, then as now the largest grain collection in the United States. There, the amazing grain—the diverse grain, the grain of Mesoamerican and Borlaugian ingenuity—was put in some boxes at room temperature and ignored. The seeds all germinated and rotted. There were no backups.

7. At one point Harrar simply told Borlaug he had to stop the work in Sonora. It wasn't going to work. It was too expensive. It was too hard. Borlaug resigned. The next day, Harrar gave in and told Borlaug he could keep working in Sonora. The resignation was ignored.

8. Fertilizer would prove key to Borlaug's vision for agriculture, and the industrial production of fertilizer was the result of war. Until World War I, nearly all fertilizer came from bird or bat guano. Then, during

the war, scientists at the German company BASF (Badische Anilin und Soda Fabrik, or, in English, Baden Aniline and Soda Factory) were tasked with finding a way to produce more ammonia for bombs. The scientists developed a method we now call the Haber-Bosch process, which could be used to produce large quantities of ammonia industrially, which in turn was used in bombs during the war. After the war, BASF used the same approach—the same equipment, even—to produce more ammonia, which was then converted into fertilizer.

9. Norin 10 stood out because of how very short it was. But short wheats from Japan had been observed as early as 1873 by Horace Capron, who headed an American delegation to Japan and brought some of the wheats back to the United States. Similar wheat had previously been distributed from Japan to France. Later, other short wheats were shared between Japan and Italy. It is tempting to see the Americans as having unjustly taken Norin 10 from the Japanese, but it is worth noting that Norin 10 itself is the result of many earlier borrowings. The Japanese had crossed a variety of wheat called Fultz, obtained from the United States (where it had been taken from the Mediterranean), with a Japanese variety called Daruma. The result of this cross was then crossed with a variety of wheat that had been brought from Russia to Kansas by Mennonite immigrants. Norin 10 would ultimately result from further work that Inazuka did on the basis of this latter cross. The free sharing of varieties was key to Borlaug's version of progress.

10. Perkins, *Geopolitics and the Green Revolution*.

11. The institute's website (http://irri.org/about-us/our-history) offers a nice timetable of its history, along with historical videos of the work at the institute.

12. It is interesting to consider the ways in which various regions contribute to the world. Borlaug spread the Green Revolution around the globe and with it a new way of farming, eating, and living that is a key aspect of modern Western life, however one feels about it. As for the French, as a different sort of example, they offered the world the restaurant. The waiter, the napkins, the number of courses, that the last course is sweet, that the first course is small, that a specific kind of wine might go with a specific kind of food—that's all French. Even the word *restaurant* refers to a sort of fortifying meat broth (with restorative powers) associated with the very venues where one might go out to eat, take a date, hint at romance, savor food—oh, really savor it—and savor, too, the conversation. In other words, Borlaug gave us sustenance, food to sustain us, and the French, they offered something else: tables and meals, and with them the restorative power of both flavors and conversations.

13. Many argue that it was this success that the funders, including the Ford Foundation, the Rockefeller Foundation, the US government, and the World Bank, were most eager to see happen. The new agriculture provided food and, it was thought, forestalled socialism at the same time. See Cary Fowler and Pat Mooney, *Shattering: Food, Politics, and the Loss of Genetic Diversity* (Tucson: University of Arizona Press, 1990).

Chapter 11

1. See Schultes's own account, "The Domestication of the Rubber Tree: Economic and Sociological Implications," *American Journal of Economics and Sociology* 52, no. 4 (October 1993): 479–85. See also Wade Davis's brilliant biogeography of Schultes and his work, *One River: Explorations and Discoveries in the Amazon Rain Forest* (New York: Simon and Schuster, 1996). Yes, the seeds could be eaten, though only after a fair amount of work. Like cassava roots, the seeds of *H. brasiliensis* contain cyanide and so had to be soaked, boiled, and soaked again in order to be made palatable and safe.

2. Wickham, whose actions the modern world is greatly dependent upon, was a character. Before he gathered rubber seeds, he had first tried to make his name and fortune in Nicaragua by shooting birds in order to pluck their feathers, which he would then sell to his mother, who ran a millinery shop in London. He failed, in no small part because of the fact that he was not very good at shooting birds. See Joe Jackson, *The Thief at the End of the World: Rubber, Power, and the Seeds of Empire* (New York: Viking, 2008).

3. Joseph was the son of William Hooker, the former director of Kew. The father and son worked together for forty years and even inspired a book entitled *The Hookers of Kew,* which sold very well in the United States, perhaps because of a misunderstanding of the title.

4. Stuart McCook and John Vandermeer, "The Big Rust and the Red Queen: Long-Term Perspectives on Coffee Rust Research," *Phytopathology* 105, no. 9 (September 2015): 1164–73.

5. Though interestingly, the biology of the trees and plantations still circumscribed the lives of the rubber tappers, even in Asia. Whereas in the Amazon a very busy tapper, working until he was nearly incapacitated each day, could tap eighty-five trees, in tropical Asian plantations a tapper could tap 350. Close planting made the travel easier, but each day was (and is) still composed of a circuit of very well known trees.

6. Which is to say that by 1921 Asia was producing ten times as much latex as was being produced in the Amazon during the peak years of extraction. Ten times.

7. Julian Street, *Abroad at Home: American Ramblings, Observations, and Adventures of Julian Street* (New York: Century Company, 1914).

8. Braz Tavares da Hora Júnior, et al., "Erasing the Past: A New Identity for the Damoclean Pathogen Causing South American Leaf Blight of Rubber," *PLoS One* 9, no. 8 (August 2014): e104750. In this study, the authors rename the fungus but also note that it had been so poorly studied genetically that our knowledge of which other species it might be related to was ambiguous. Worse yet, our knowledge of which other species are really one and the same with it was also ambiguous. It was only through this very recent work that the life history of the fungus was fully resolved and that we realized this fungus's close relation to other terrible plant pathogens, including the one that causes black Sigatoka, a disease affecting bananas.

9. At this point, several scientists had already warned of the dangers of leaf blight to rubber plantations, but apparently neither Ford nor those who worked with him read W. N. C. Belgrave, "Notes on the South American Leaf Disease of Rubber," *Journal of the Board of Agriculture of British Guiana* 15 (1922): 132–38, or J. R. Weir, "The South American Leaf Blight and Disease Resistant Rubber," *Quarterly Journal of the Rubber Research Institute of Malaya* 1 (1929): 91–97. Kids, this is why we have to publish in open-access journals.

10. It was around $20 million in 1940s dollars.

11. It was also a dream of Adolf Hitler. Hitler went so far as to announce in 1935, at the seventh Nazi party congress, in Nuremberg, that the Nazis had figured out how to produce synthetic rubber. Fortunately, he was lying.

12. The Americans not only saw the need to be able to produce their own rubber, they also saw the extent to which Europeans could be dependent on it. As a result, rubber production factories in both Germany and Italy were among the most important bombing targets during the war.

13. There are only eleven known species of *Hevea*, which is to say that Schultes was able to collect, on his personal expedition, all but one of the *Hevea* species known in the world. See Reinhard Lieberei, "South American Leaf Blight of the Rubber Tree (*Hevea* spp.): New Steps in Plant Domestication using Physiological Features and Molecular Markers," *Annals of Botany* 100, no. 6 (December 2007), 1125–42, at http://www.ncbi.nlm.nih.gov/pmc/articles/PMC2759241/pdf/mcm133.pdf.

14. As Wade Davis notes in his book *One River: Explorations and Discoveries in the Amazon Rain Forest* (New York, Simon and Schuster, 1996), a history of both rubber exploration and Schultes.

15. Davis, *One River,* 537.

NOTES

16. Daniel C. Ilut, et al., "Genomic Diversity and Phylogenetic Relationships in the Genus *Parthenium* (Asteraceae)," *Industrial Crops and Products* 76 (2015): 920–29.

Chapter 12

1. Oghenekome Onokpise and Clifford Louime, "The Potential of the South American Leaf Blight as a Biological Agent," *Sustainability* 4, no. 11 (November 2012): 3151–57, http://www.mdpi.com/2071-1050/4/11/3151.
2. Kheng Hoy Chee, "Management of South American Leaf Blight," *The Planter* 56 (1980), 314–25.
3. One other solution, although we would need to start planning a decade or more before we would expect to see results, would be to plant rubber in those places where rubber trees grow but the blight cannot. In essence, the approach would be to move the rubber not so much geographically as climatically—to extreme climates. See Franck Rivano, et al., "Suitable Rubber Growing in Ecuador: An Approach to South American Leaf Blight," *Industrial Crops and Products* 66 (April 2015): 262–70. Yet while this seems promising and appears to be working on a small scale in Guatemala, Ecuador, and parts of Brazil too extreme in climate for the blight, it would need to happen now in order to be able to save rubber in fifteen years. In addition, the odds that leaf blight would evolve to colonize new climates, if many rubber trees were there, seem high.
4. Hans ter Steege, et al., "Estimating the Global Conservation Status of More Than 15,000 Amazonian Tree Species," *Science Advances* 1, no. 10 (November 2015): e1500936.
5. Even if more resistant trees are produced, if they do not yield as much latex as the trees planted in Asia, none is likely to be planted at any scale in Asia unless leaf blight arrives. At that point, any efforts to replant with resistant varieties on a large scale would take time (for trees to grow) and would require additional land in order to yield the same amount of rubber already being produced. In theory, the trees that have been planted experimentally in Brazil and elsewhere in the Americas might next be grown in greater numbers. New plantations like those Ford established might be grown. But when they are, they will be less productive than the plantations in Asia (and thus likely less profitable). In addition, farmers will deal with leaf blight as a certainty rather than a possibility. What's more, as leaf blight has begun to be better studied (although still far less well than one might hope), it has been revealed to be both very diverse and rapidly evolving. In other words, many leaf blights exist, with varying abilities to deal with resistant trees. That some of these varieties of leaf blight might be able to destroy even the resistant trees is, well, possible.

293

6. Gary Paul Nabhan, *Where Our Food Comes From: Retracing Nikolay Vavilov's Quest to End Famine* (Washington, DC: Island Press, 2008).

7. Francesco Emanuelli, et al., "Genetic Diversity and Population Structure Assessed by SSR and SNP Markers in a Large Germplasm Collection of Grape," *BMC Plant Biology* 13 (2013): 39, at http://bmcplantbiol .biomedcentral.com/articles/10.1186/1471-2229-13-39. The challenge of our dependence upon the occasional reproduction of traditional varieties of crops and their wild relatives is that it is most likely to occur in the places where the traditional varieties of crops are hardest to farm—in their native ranges, where their pests and pathogens are most diverse. As a result, such farms have far greater value than what might be suggested by the cost of their products in the market. The value of traditional varieties of corn being farmed in Mexico, for example, is not only to local culture and conservation. It is also—because Mexico is one of relatively few places where the genes of wild teosinte might occasionally mix with those of traditional corn and produce new varieties—a boon to corn planted anywhere on earth. It benefits us all when we subsidize the farming of traditional varieties of crops in their native ranges, whether through government subsidies or the extra money we pay to buy such crops.

8. Squash bees are the main pollinator of squash plants, or at least they are in the places in which squash are native or were moved before the time of conquistadors. See Margarita M. López-Uribe, et al., "Crop Domestication Facilitated Rapid Geographical Expansion of a Specialist Pollinator, the Squash Bee *Peponapis pruinosa*," *Proceedings of the Royal Society B: Biological Sciences* 283, no. 1833 (June 2016). We know relatively little about the pollination of the squash plants that have been moved on ships from the Americas to Africa and Asia. Similarly, little work has been done to understand how the relative of squash, the bottle gourd, is pollinated where it has been introduced. Or, for that matter, how cucumbers (distant kin to the squash) are pollinated in the Americas far from the pollinators with which they evolved. Our general tendency is to move our crops from place to place and hope that somehow these wonderful mutualists come along, or that their roles continue to be carried out by some local replacement, an ecological stand-in. Sometimes this works. Nature forgives some of our faults. In other cases, we figure out after many years of struggle that something quite important has been left behind.

9. Though it is worth noting that even resistance derived from a crop's wild relatives is not permanent. Pests and pathogens continue to evolve. A strain of grassy stunt virus, for example, has been found to have

evolved the ability to overcome the resistance to it bred from wild rice into domesticated rice.

10. Brian V. Ford-Lloyd, et al., "Crop Wild Relatives—Undervalued, Underutilized and Under Threat?," *BioScience* 61, no. 7 (2011): 559–65.

11. Nigel Maxted, et al., "Toward the Systematic Conservation of Global Crop Wild Relative Diversity," *Crop Science* 52, no. 2 (2012): 774–85.

12. Hallie Eakin, "Institutional Change, Climate Risk, and Rural Vulnerability: Cases from Central Mexico," *World Development* 33, no. 11 (November 2005): 1923–38.

13. Tsutomu Ishimaru, et al., "A Genetic Resource for Early-Morning Flowering Trait of Wild Rice *Oryza officinalis* to Mitigate High Temperature-Induced Spikelet Sterility at Anthesis," *Annals of Botany* 106 no. 3 (2010): 515–20. See also Hannes Dempewolf, et al., "Adapting Agriculture to Climate Change: A Global Initiative to Collect, Conserve, and Use Crop Wild Relatives," *Agroecology and Sustainable Food Systems* 38, no. 4 (2014): 369–77.

14. Neil Brummitt and Steven Bachman, *Plants Under Pressure: A Global Assessment—The First Report of the IUCN Sampled Red List Index for Plants* (London: Royal Botanic Gardens, Kew, 2010). Also see Colin K. Khoury, et al., "An Inventory of Crop Wild Relatives of the United States," *Crop Science* 53, no. 4 (2013): 1496–1508 for more about crop wild relatives in the specific context of the United States.

15. Cade Metz, "The Superplant That May Finally Topple the Rubber Monopoly," *Wired,* July 13, 2015, at http://www.wired.com/2015/07/superplant-may-finally-topple-rubber-monopoly/.

16. Ibid.

17. The Aztecs used guayule to produce rubber. The word *guayule* derives from the Nahuatl word *ulli,* for "rubber"; the plant was one of two used by the Aztecs for rubber production. The rubber balls the Aztecs used in their sports may have been produced from guayule. Beginning in the early 1900s, the United States began to harvest guayule in Arizona and then imported it from Mexico for use in tires. It was gathered as a buffer against the worry that most of the supply of rubber had been monopolized by Asia. By 1910, roughly 10 percent of the rubber in the world came from guayule. But there was a problem. Wild populations of the plant were quickly disappearing. By 1911, perhaps as little as 20 percent of the populations of the plant present in 1890 in Mexico and the United States remained. Investors, including the Rockefeller Foundation, decided in the early 1900s (even before Ford dreamed of a Brazilian empire) to try to farm guayule to produce rubber. They invested $30

million in the project, which built upon attempts at the domestication of guayule that had begun in 1910. The attempt moved forward, but slowly. Then in the 1940s, during World War II, tens of thousands of acres were planted in guayule. The guayule grew, but not quickly enough. By the time things were starting to take off, the war ended, artificial rubber was invented, and the perceived need for an alternative rubber source evaporated.

18. Charles Darwin, *On the Origin of Species by Means of Natural Selection: Or the Preservation of Favoured Races in the Struggle for Life*, 4th ed. (London: John Murray, 1866).

19. For example, the wild rice *Oryza rufipogon* has been shown to support seven times as many species as domesticated rice, *O. sativa*. See Yolanda H. Chen, Rieta Gols, and Betty Benrey, "Crop Domestication and Its Impact on Naturally Selected Trophic Interactions," *Annual Review of Entomology* 60 (January 2015): 35–58.

20. There are some exceptions to this generalization. War is full of exceptions, contradictions, caveats, and complexities, one of them being that in some cases the no-man's-land between warring tribes, clans, or regions can be an area in which conservation occurs by default because of the relative absence of humans. The area, for example, in which Lewis and Clark saw bison in the greatest abundance appears to have been just such a no-man's-land: the warring tribes were unwilling to enter that region. See Paul S. Martin and Christine R. Szuter, "War Zones and Game Sinks in Lewis and Clark's West," *Conservation Biology* 13, no. 1 (February 1999): 36–45. The same may be true for some crop wild relatives.

Chapter 13

1. The lesson I learned in this particular context was to never ask, "Hey, what's in that jar on your shelf?"

2. Kristen Mack, "Leigh Van Valen, 1935–2010: Evolutionary Biologist Coined 'Red Queen' Theory of Extinction," *Chicago Tribune*, October 24, 2010, at http://articles.chicagotribune.com/2010-10-24/features/ct-met -obit-van-valen-20101024_1_extinction-journals-modern-biology. Generally speaking, the word *quirky* is a euphemism for some better and more interesting but slightly offensive description.

3. Comparison from Matt Ridley, *The Red Queen: Sex and the Evolution of Human Nature* (London: Penguin Books, 1994). Though if we are going to get picky, Van Valen's beard seems as though it was just about exactly the same length as that of Leonardo da Vinci.

4. For the interested and the bold, "Mating Behavior in the Dinosauria": https://dl.dropboxusercontent.com/u/18310184/songs/dino-wedding.pdf.

5. Mack, "Leigh Van Valen."
6. He was voted "most academic" in the second grade. In the second grade!
7. And he was rarely stultified. Van Valen once named, for example, twenty fossil mammals after characters in J. R. R. Tolkien's fiction. See Douglas Martin, "Leigh van Valen, Evolution Revolutionary, Dies at 76," *New York Times,* October 30, 2010, at http://www.nytimes.com/2010/10/31/us/31valen.html.
8. Leigh Van Valen, "A New Evolutionary Law," *Evolutionary Theory* 1 (1973): 1–30. *Evolutionary Theory* would go on to be published sporadically for years and attracted to its editorial board some of the greatest luminaries of evolutionary biology.
9. John F. Tooker and Steven D. Frank, "Genotypically Diverse Cultivar Mixtures for Insect Pest Management and Increased Crop Yields," *Journal of Applied Ecology* 49, no. 5 (October 2012): 974–85. Evidence also suggests that diversity of crop species in or adjacent to a field may also reduce crop loss in the first place, even before a shift is made from less to more resistant varieties. However, the extent of this benefit seems variable. See Deborah K. Letourneau, et al., "Does Plant Diversity Benefit Agroecosystems? A Synthetic Review," *Ecological Applications* 21, no. 1 (January 2011): 9–21. Traditional farms also vary greatly, from culture to culture and region to region, in the ways in which crops are farmed. Most of this variation has been poorly studied, even more so than the seeds themselves. As a result, while the knowledge embedded in traditional farming systems around the world is potentially very useful as we think about the future of crops, is rapidly being lost.
10. Jelle Bruinsma, ed., *World Agriculture: Towards 2015/2020: An FAO Perspective* (Rome, Italy: Food and Agriculture Organization of the United Nations, 2003).
11. The consequences are predictable not only in general but also in many details (though nature doesn't always perfectly heed predictions). For example, theory predicts that the more tropical the region in which one of these crops is planted, the more quickly resistance may evolve. This might occur for two reasons. First, the tropics are much more diverse than are temperate regions, so the number of insect, fungus, and oomycete species that land in fields and try to eat crops is much higher. In the lottery of species, having more species try to win and having them try to win more times per year pays off, at least for them. Second, in the tropics, generation times are faster, and so because biological stories are measured in generations, not years, time fast-forwards in proportion to the number of generations pests and pathogens produce per year. Though this second explanation does come with a caveat. If generations are very short in

insects, they may fail to be exposed to kill them (and their populations hence suffer little selection). If generations are very long, all the insects in the population may die (and hence suffer no selection, since no individuals are left over to be favored). It may be that in these cases that organisms with intermediate generation times evolve most rapidly, as has been suggested for existing examples of resistance. See for example Rosenheim, Jay A., and Bruce E. Tabashnik. "Influence of generation time on the rate of response to selection." *American Naturalist* (1991): 527–541.

12. Andrew J. Forgash, "History, Evolution, and Consequences of Insecticide Resistance," *Pesticide Biochemistry and Physiology* 22, no. 2 (October 1984): 178–86.

13. University breeding programs and public breeding programs have been gutted in direct proportion to the success of the Green Revolution.

14. Youyong Zhu, et al., "Genetic Diversity and Disease Control in Rice," *Nature* 406 (August 17, 2000): 718–22.

15. We are studying the microbes of household insects in my lab to see if they can turn paper waste into energy.

16. Bart Lambert and Marnix Peferoen, "Insecticidal Promise of *Bacillus thuringiensis*," *BioScience* 42, no. 2 (February 1992): 112–22.

17. Richard J. Milner, "History of *Bacillus thuringiensis*," *Agriculture, Ecosystems & Environment* 49, no. 1 (May 1994): 9–13.

18. Though it is worth nothing that these are the same minds, the same beautifully focused minds, about which one could easily write many comedic novels—beautiful minds peering deep into the mysteries of life but coming up only rarely to get a sense of context.

19. Tina Kyndt, et al., "The Genome of Cultivated Sweet Potato Contains *Agrobacterium* T-DNAs with Expressed Genes: An Example of a Naturally Transgenic Food Crop," *Proceedings of the National Academy of Sciences of the United States of America* 112, no. 18 (May 5, 2015): 5844–49.

20. Though this is a gradient, not a set of categories. It is probably more sensible to talk about forms of genetic engineering imposed by humans on crops. At one extreme is the Native American farmer who chooses corn with sugary stalks over those with less sugary stalks, and in doing so engenders/creates/engineers a crop that has the genes associated with a sugary stalk. At the opposite extreme is the activity of engineering plants that produce human breast milk. Engineering a plant to produce a toxin it didn't already produce is somewhere between these extremes.

Chapter 14

1. John Seabrook, "Sowing for Apocalypse: The Quest for a Global Seed Bank," *New Yorker,* August 27, 2007.

2. Frances Moore Lappé and Joseph Collins with Cary Fowler, *Food First: Beyond the Myth of Scarcity* (Boston: Houghton Mifflin, 1977). Lappé was by then already well known as author of the bestselling *Diet for a Small Planet,* which advocated vegetarianism as an ecologically sustainable lifestyle.
3. Harlan was by no means the first to voice this argument. Erwin Baur, a German botanist, had already noticed the shift as early as 1914. See his "Die Bedeutung der primitiven Kulturrassen und der wilden Verwandten unserer Kulturpflanzen für die Pflanzenzüchtung" [The importance of the primitive cultivars and wild relatives of our crops for plant breeding], *Jahrbuch der Deutschen Landwirtschafts Gesellschaft* 29 (1914): 104–10.
4. Jack R. Harlan, "Genetics of Disaster," *Journal of Environmental Quality* 1, no. 3 (July–September 1972): 212–15.
5. Jack Harlan was quick to credit others who had voiced similar sentiments before him, but because the man he gave the most credit to was Harry Harlan, his dad, who was a friend of Vavilov, we can still call this hypothesis Harlan's ratchet while acknowledging some ambiguity as to which Harlan we are talking about.
6. Alfred W. Crosby Jr., *The Columbian Exchange: Biological and Cultural Consequences of 1492* (Westport, CT: Praeger Publishers, 2003).
7. Cary Fowler and Pat Mooney, *Shattering: Food, Politics, and the Loss of Genetic Diversity* (Tucson: University of Arizona Press, 1990).
8. Seabrook, "Sowing for Apocalypse."
9. This assumes that the genes for resistance are recessive. If they are dominant, the refuge strategy would still delay resistance, but not forever.
10. For a thoughtful review of what was and was not working in the first billion acres of Bt crops, see Bruce E. Tabashnik, Thierry Brévault, and Yves Carrière, "Insect Resistance to Bt Crops: Lessons from the First Billion Acres," *Nature Biotechnology* 31, no. 6 (June 2013): 510–21.
11. Kong-Ming Wu, et al., "Suppression of Cotton Bollworm in Multiple Crops in China in Areas with Bt Toxin–Containing Cotton," *Science* 321, no. 5896 (September 19, 2008): 1676–78; Janet E. Carpenter, "Peer-Reviewed Surveys Indicate Positive Impact of Commercialized GM Crops," *Nature Biotechnology* 28 (2010): 319–21; William D. Hutchison, et al., "Areawide Suppression of European Corn Borer with Bt Maize Reaps Savings to Non-Bt Maize Growers," *Science* 330, no. 6001 (October 8, 2010): 222–25; Michael D. Edgerton, et al., "Transgenic Insect Resistance Traits Increase Corn Yield and Yield Stability," *Nature Biotechnology* 30 (2012): 493–96; Jonas Kathage and Matin Qaim, "Economic Impacts and Impact Dynamics of Bt (*Bacillus thuringiensis*) Cotton

in India," *Proceedings of the National Academy of Sciences of the United States of America* 109, no. 29 (July 17, 2012): 11652–56.

12. Yanhui Lu, et al., "Widespread Adoption of Bt Cotton and Insecticide Decrease Promotes Biocontrol Services," *Nature* 487 (July 19, 2012): 362–65.

13. Bruce E. Tabashnik and Yves Carrière, "Successes and Failures of Transgenic Bt Crops: Global Patterns of Field-Evolved Resistance," in *Bt Resistance: Characterization and Strategies for GM Crops Producing Bacillus Thuringiensis Toxins*, ed. Mario Soberon, Yulin Gao, and Alejandra Bravo (Boston: Centre for Agriculture and Biosciences International, 2015): 1–4.

14. In part for this reason, some have argued for policies that require companies to pull their transgenic crops off of the market once resistance emerges. However, this approach would work best if all countries, or at least all countries in a region, comply, especially in regions in which organisms can move easily across international borders.

15. As Fowler gathered copies of seeds from around the world, he knew he would need copies of Vavilov's seeds. John Seabrook reports in his *New Yorker* article that when Fowler went to Vavilov's collection, he did not know what he would find. Some said the collections still contained seeds; others said that many of the seeds had died and that much of what Vavilov gathered had gone literally to rot. To get to the seeds, Seabrook and Fowler, together, met with Nikolai Dzubenko, the man who now sits in Vavilov's office and maintains his legacy. They ate with him in a room in the same building where Vavilov's seed savers died during the siege. They ate "pastries, fruits, cold meats, cheeses, juice," and drank vodka. On that day, Fowler, who was sick, toured the new storage facilities of the VIR, where things were shiny but not yet functional. So, by reports, they remain.

16. The headquarters of the Nordic Gene Bank is in Alnarp, Sweden.

17. Via a donation to the Global Crop Diversity Trust. This donation also supported work to produce extra copies of seeds in the seed banks of developing countries and then to ship the extras to Svalbard.

18. The perception both in the United States and Europe is that transgenic crops are less healthful than conventional crops. A study conducted by the Pew Research Center in 2015 found that 57 percent of Americans thought it was harmful to eat genetically modified food. Yet so far evidence of health problems associated with consuming transgenic crops is lacking. The National Academy of Sciences, the American Medical Association, the American Association for the Advancement of Science, and the World Health Organization have concluded that products made from transgenic crops now on the market are safe to eat. Of course the transgenic crop we are talking about is, mostly, corn—corn that is being used,

in part, to produce high-fructose corn syrup. No one debates the fact that high fructose corn syrup is bad for you—it is—but its negative impact has little to do with whether the corn is transgenic. We should continue to be cautious about each new thing we consume, but nothing about the process of making a transgenic crop, as opposed to breeding a crop via more conventional means, poses a categorically new risk.

19. Transgenic crops with Bt genes might, for example, have negative effects on beneficial or rare insects. Monarch butterflies in particular have been the subject of concern. Early on, one study demonstrated Bt crops' negative effect on monarch butterflies (via pollen). A series of subsequent studies has, however, found no effect on monarch butterflies, whether at concentrations of Bt toxins like those actually found in fields or at even higher concentrations. Other studies show that many beneficial insects (those that eat pests) do better on Bt crops than they do on Green Revolution varieties sprayed with pesticides, because the pesticides kill the beneficial insects as well as the Bt crops' target pests. Crops that are engineered to be resistant to herbicides, on the other hand, seem to be a different story. These crops lead farmers to spray large quantities of herbicides that spill over into adjacent habitats and river systems, killing plants and polluting waterways and aquifers. Each transgenic crop is likely to be different in its consequences. See Emily Waltz, "GM Crops: Battlefield," *Nature* 461 (September 3, 2009): 27–32; Mike Mendelsohn, et al., "Are *Bt* Crops Safe?", *Nature Biotechnology* 21, no. 9 (September 2003): 1003–9, doi:10.1038/nbt0903-1003; Richard L. Hellmich, et al., "Monarch Larvae Sensitivity to *Bacillus thuringiensis*–Purified Proteins and Pollen," *Proceedings of the National Academy of Sciences of the United States of America* 98, no. 21 (October 9, 2001): 11925–30.

20. The risk of unknown consequences is a reasonable one when confronted with new technologies. It was reasonable in the early days of heart transplants, when transplants were performed before they were safe or beneficial. It was reasonable in the early days of nuclear power. Urging caution upon the progress of technology has, historically, often been justified. There is a reason Mary Shelley's *Frankenstein* still resonates. In the cases of transgenic crops, the worry relates to whether their genes might spread to other organisms. The worry relates to unknown things they might do to our bodies. The worry relates to the unknown. The worry should apply to any crop, just as it should apply to any pesticide, even those used on organic crops (which, ironically, is where Bt was first sprayed and is still used).

21. Jonathan Knight, "Crop Improvement: A Dying Breed," *Nature* 421 (February 6, 2003): 568–70.

22. Kenneth J. Frey, *National Plant Breeding Study—I: Human and Financial Resources Devoted to Plant Breeding Research and Development in the United States in 1994* (Ames, IA: Iowa State University, Iowa Agriculture and Home Economics Experiment Station, 1996).

23. Virginia Gewin, "New Film Traces Cary Fowler's Quest to Build the Doomsday Seed Vault," *Science,* May 15, 2015, at http://www.sciencemag .org/news/2015/05/new-film-traces-cary-fowler-s-quest-build-doomsday -seed-vault.

24. Knight, "Crop Improvement."

25. David Greene, "Researchers Fight to Save Fruits of Their Labor," *Morning Edition,* August 30, 2010, at http://www.npr.org/templates/ story/story.php?storyId=129499099.

26. Ibid.

27. Ibid.

28. Tom Parfitt, "Pavlovsk's Hopes Hang on a Tweet," *Science* 329 (August 20, 2010): 899.

29. See the order at http://www.vir.nw.ru/news/14.05.2012_en.html.

30. See, for example, DivSeek: http://www.divseek.org/mission-and-goals/.

Chapter 15

1. Some can be found in Vavilov's collection, potentially including those collected by Vavilov himself. Others are in the USDA's National Small Grains Collection. Others still are in the ICARDA collection, in Syria, or the doomsday vault, in Norway.

2. Tom Clarke, "Seed Bank Raises Hope of Iraqi Crop Comeback," *Nature* 424 (July 17, 2003): 242.

3. Glen R. Gibson, James B. Campbell, and Randolph H. Wynne, "Three Decades of War and Food Insecurity in Iraq," *Photogrammetric Engineering and Remote Sensing* 78, no. 8 (August 2012): 885–95.

4. Issued by Paul Bremer, then head of the Coalition Provisional Authority.

5. Or, as some have suggested, maybe the change came about because of beer. Several authors have now suggested that even before agriculture beer was made from sprouted grain (particularly barley) left to rot. The slurry was consumed. It was enjoyed. Soon so much grain was being gathered to make beer that not enough was left over to eat. More was needed. Alcohol may have figured similarly in the earliest days of domestication of corn in the Americas and rice in Asia. For more on the settlements in which grains were first gathered see Riehl, Simone, Mohsen Zeidi, and Nicholas J. Conard. "Emergence of agriculture in the foothills of the Zagros Mountains of Iran." *Science* 341.6141 (2013): 65–67. For more on the role of climate in the subsequent transi-

tion to active agriculture see Borrell, Ferran, Aripekka Junno, and Joan Antón Barceló. "Synchronous environmental and cultural change in the emergence of agricultural economies 10,000 years ago in the Levant." *PloS one* 10.8 (2015): e0134810.

6. In many cases new traits were favored in seeds by farmers passively. That is, farmers did not necessarily choose these traits as much as the process of planting, gathering, and storing seeds favored some traits over others. Nonshattering, for example, has been argued to have emerged in some grains as a chance mutation, one that, as Greg Anderson noted in an e-mail to me, is bad for the plants. The plants with grass seeds (which are really a kind of tiny fruit, a caryopsis) that shattered were more likely to be dispersed in the field, either because the shattered seeds blew in the wind or rode on mammals. But those that did not shatter, the rare few with a mutation that prevented shattering, were more likely to be gathered, more likely to be stored, and more likely to be planted anew the next year. Farmers didn't search out certain genes. Instead they collected the seeds that hadn't fallen off the plant and thus favored the genes for seeds that didn't shatter, that didn't fall.

7. Where *durum* just means hard. The grains of durum wheat are relatively hard to grind.

8. Durum wheat appears to be a cross between a grass called einkorn and another grass—we don't yet know which one. This cross doubled the number of copies of chromosomes durum wheat has relative to either ancestor. Durum wheat was then later crossed with another grass, goat grass, and produced a wheat variety with two more sets of chromosomes, a variety that gave rise to bread wheat. Most wheat grown today is bread wheat.

9. It was a realm complete with names, so recent is our civilized history. The first ruler of this empire was Sargon of Akkad. Independent cities (Uruk, Ur, Umma, Kish) preceded the empire over which Akkad ruled. He united the cities and created, in its grandest sense, Babylonia. It was an empire we would recognize in many respects.

10. Most accounts now suggest that early agriculture was so desperate that, with its dawn, life expectancies decreased. Even where the total number of calories available to a place increased because of agriculture, the nutritional value of those calories decreased, and the social inequalities of their distribution increased (at least initially). Pathogens became common. Bones found at archaeological sites from this time bear the pocks of malnutrition; teeth, for the first time, frequently bore signs of abscesses.

11. Ahmed Amri, et al., "Chromosomal Location of the Hessian Fly Resistance Gene *H20* in 'Jori' Durum Wheat," *Journal of Heredity* 81, no. 1

(1990): 71. Ahmed Amri, et al., "Complementary Action of Genes for Hessian Fly Resistance in the Wheat Cultivar 'Seneca,'" *Journal of Heredity* 81, no. 3 (1990): 224–27; Ahmed Amri, et al., "Resistance to Hessian Fly from North African Durum Wheat Germplasm," *Crop Science* 30, no. 2 (1990): 378–81.

12. ICARDA, as a reminder, is one of fifteen similar centers, each with a specific goal. Each center focuses on saving seeds as well as on working with farmers to save local farming practices. One center focuses on rice—the International Rice Research Institute, in the Philippines. Then there is the International Potato Center, in Peru, about which I have written (see page 198). Two of these centers, ICARDA and CIMMYT, both have a focus on wheat. But whereas CIMMYT focuses on Green Revolution wheats, which grow well with lots of water and fertilizer, the varieties ICARDA works on do the opposite: they do well with little water and little in the way of chemical inputs.

13. This variety alone had a huge effect in Syria, reducing by half the area that needed to be sprayed with pesticides against sunn pests.

14. Akia Kitoh, et al., "First Super-High-Resolution Model Projection That the Ancient 'Fertile Crescent' Will Disappear in This Century," *Hydrological Research Letters* 2 (2008): 1–4, doi:10.3178/HRL.2.1.

15. We are eager to believe ourselves, as humans, too clever to succumb to simple tragedies such as those posed by climate. The archaeologist Joseph Tainter said, for instance (wrongly, as it turns out), that it is doubtful that "any large society has ever succumbed to a single-event catastrophe." See Joseph A. Tainter, *The Collapse of Complex Societies* (Cambridge, UK: Cambridge University Press, 1990).

16. Alexia Smith, "Akkadian and Post-Akkadian Plant Use at Tell Leilan," in *Seven Generations Since the Fall of Akkad,* ed. Harvey Weiss (Wiesbaden: Harrassowitz Verlag, 2012): 225–40.

17. Harvey Weiss, et al., "The Genesis and Collapse of Third Millennium North Mesopotamian Civilization," *Science* 261, no. 5124 (August 20, 1993): 995–1004.

18. Pinhas Alpert and Jehuda Neumann, "An Ancient 'Correlation' Between Streamflow and Distant Rainfall in the Near East," *Journal of Near Eastern Studies* 48, no. 4 (October 1989): 313–14.

19. Hugh Garnet McKenzie, "Skeletal Evidence for Health and Disease at Bronze Age Tell Leilan, Syria" (master's thesis, University of Alberta, 1999).

20. For more on this amazing history, see Elizabeth Kolbert, "The Curse of Akkad," in *Field Notes from a Catastrophe: Man, Nature, and Climate Change* (New York: Bloomsbury USA, 2006), and Harvey Weiss, "Late Third Millennium Abrupt Climate Change and Social Collapse

in West Asia and Egypt," in *Third Millennium BC Climate Change and Old World Collapse*, ed. H. Nüzhet Dalfes, George Kukla, and Harvey Weiss (Berlin, Heidelberg, and New York: Springer International Publishing, 1997), 711–23.

21. Editorial, "Seeds in Threatened Soil," *Nature* 435 (June 2, 2005): 537–38, doi:10.1038/435537b.

22. Iraq faced the same drought, but it did so without the accompanying drawdown of groundwater. In addition, much of the food being consumed in Iraq came during those years—and still comes—from subsidized imports from the United States and Europe, imported grains cheaper to buy than to grow. My reconstruction of the drought and its consequences in Syria is based heavily on the excellent work of Colin Kelley and his colleagues. See Colin P. Kelley, et al., "Climate Change in the Fertile Crescent and Implications of the Recent Syrian Drought," *Proceedings of the National Academy of Sciences* 112, no. 11 (March 17, 2015): 3241–46.

23. The trust was set up by CGIAR and the United Nations' Food and Agriculture Organization (FAO).

24. United States Department of Agriculture, Foreign Agricultural Service, Commodity Intelligence Report, "Syria: Wheat Production in 2008/09 Declines Owing to Season-Long Drought," at http://www.pecad.fas.usda.gov/highlights/2008/05/Syria_may2008.htm.

25. And, frankly, so was much of the Middle East. In the broader region described as CWANA—Central and West Asia and North Africa, a region that includes a billion people—only Kazakhstan, Turkey, and Syria produced enough wheat before 2011 to feed their people. Now, just Kazakhstan and Turkey do. Wheat is key to the story of food production and consumption in this region, both because it is the most consumed crop (by far, comprising 37 percent of calories) and because it is the only crop whose yield and production have been increasing in the region as populations have increased.

26. Anthony Sattin, "Syria Burning: ISIS and the Death of the Arab Spring Review—How a Small-Scale Revolt Descended into Hell," *The Guardian*, August 9, 2015, at http://www.theguardian.com/books/2015/aug/09/syria-burning-isis-death-arab-spring-review-revolt-hell.

27. The research on characterization and multiplication of genetic resources and on variety development were relocated to the research platforms established in Lebanon, Ethiopia, India, and Morocco. There, workers took measurements of yields, yields that would be important when the war was over, so that the very best crops could be shared with farmers in Syria and beyond.

28. Hazem Badr, "Syria's ICARDA Falls to Rebels, but Research Goes On," *SciDevNet,* May 22, 2014, at http://www.scidev.net/global/r-d/news/syria-s-icarda-falls-to-rebels-but-research-goes-on.html.

29. Virginia Gewin, "Crop Seed Banks Are Preserving the Future of Agriculture. But Who's Preserving Them?," *Policy Innovations,* November 19, 2015, at http://www.policyinnovations.org/ideas/commentary/data/00400.

30. The statement is in a press release from ICARDA dated September 26, 2015: https://www.icarda.org/update/press-release-icarda-safeguards -world-heritage-genetic-resources-during-conflict-syria#sthash .LoKw8vkc.dpbs.

31. Jason Ur, "Urban Adaptations to Climate Change in Northern Mesopotamia," in *Climate and Ancient Societies,* ed. Susanne Kerner, Rachael Dann, and Pernille Bangsgaard (Copenhagen: Museum Tusculanum Press, 2015): 69. In this paper Ur reconstructs shifts in the approach to agriculture in two fascinating ways. First he considers the extent of the intensification of agriculture based on the density of potsherds in fields and the number of fields containing potsherds. Potsherds are interpreted to be the residue of an approach to the fertilization (manuring) of fields that included waste of several kinds, of which the potsherds are all that remains. In addition, he reconstructs the extensification of agriculture (the proportion of the landscape farmed) by looking at what he calls hollow ways but that one might also describe as sunken trails. The trails were dug into the landscape by foot traffic. They exist where foot traffic was particularly intense. Ur interprets the intensification of foot traffic and presence of sunken trails as evidence that the landscape between the trails had become so filled with agriculture that any other route would be difficult. For an alternate (and perhaps less hopeful) interpretation of the same sites see Weiss, Harvey. "Quantifying collapse: The late third millennium Khabur Plains." Seven generations since the fall of Akkad: Wiesbaden, Harrassowitz Verlag (2012): 1–24.

32. The only change, aside from a decline in health and well-being, that seems to have preceded the collapse of Leilan was not a change in agriculture. It was rather the investment in larger and more religious buildings. As things fell apart, they begged more of their gods. This proved to be an insufficient way of dealing with climate change.

33. In the spring of 1988 a rural farmer in Uganda saw something strange on his wheat. This wheat had spots, bumps, and eruptions along its stem. Each one, upon closer examination, appeared to be releasing a rusty powder. The powder, the farmer well knew, having watched wheat for decades, was the spores of a fungus called a rust. That it was on the stem meant it was a stem rust.

Among farmers, stem rusts are a known evil. But they are an evil that had already been vanquished. The wheat growing in the field where the rust was found was resistant to rust. It was grown especially for this trait. It was not the tastiest wheat. It was not the fastest-growing wheat. But it was the wheat that would grow anywhere in the world without problems, its short stems clean and unblemished. It was impossible that the wheat growing in the farmer's field had rust.

But nature can make the impossible happen. The impossible is natural selection's ancient expertise. Natural selection produced the miniature elephant and the giant komodo lizard. It produced a hummingbird with a beak longer than its body and an isopod that eats the tongues of fish, and then, having removed them, sits in their place. Natural selection produced the wasp that flies from house to house, laying its eggs in the bodies of roaches. It is an infinitely creative carpenter, an amoral surrealist out among the flowers and leaves. Certainly this artist could also produce a rust able to live on rust-free wheat—which is just what appeared to have happened.

Chapter 16

1. As I reread the story about the preservation of smallpox, it seemed more ominous and premonitory than I remembered. The virus that causes smallpox is, indeed, kept at only two sites, one in a lab at the Centers for Disease Control and Prevention, in the United States, and another at the State Research Center of Virology and Biotechnology, in Koltsovo, Russia. The part of the story I did not know is the tragic preamble to this situation.

   The last death caused by smallpox actually occurred *after* the disease was eradicated. A medical photographer, Janet Parker, was developing pictures in her darkroom at the University of Birmingham Medical School, in England. The people in the research lab on the floor below her darkroom were conducting research on smallpox. The smallpox virus from the lab below got into the vents and traveled through them to her darkroom, where she became infected. In response, the head of the department of microbiology, Henry Bedson, committed suicide. Smallpox research stopped, but the samples in the United States and Russia were saved. It was only many years later that other samples of smallpox started to turn up—samples akin to those in the collection of plant pathogens I visited. A sample of a smallpox scab was found in a book on Civil War medicine in Santa Fe, New Mexico. Later, several vials of the smallpox virus were discovered in a lab at the National Institutes of Health. Do other vials lurk elsewhere? Almost certainly.

2. Nelson G. Hairston, Frederick E. Smith, and Lawrence Slobodkin, "Community Structure, Population Control, and Competition," *American Naturalist* 94, no. 879 (November–December 1960): 421–25. See also Lawrence B. Slobodkin, Frederick E. Smith, and Nelson G. Hairston, "Regulation in Terrestrial Ecosystems, and the Implied Balance of Nature," *American Naturalist* 101, no. 918 (March–April 1967): 109–24.

3. This was most likely a fungus that consumes dead plants rather than living ones, a wood-decay fungus, but nonetheless its growth was a reminder of the less conspicuous and yet still metabolizing life all around me in the room.

4. Elsa Youngsteadt, et al., "Do Cities Simulate Climate Change? A Comparison of Herbivore Response to Urban and Global Warming," *Global Change Biology* 21, no. 1 (January 2015): 97–105.

5. Or, given that we are talking about collections of pathogens and insects, maybe he should ride in on a giant white horsefly.

6. United Nations Department of Economic and Social Affairs, Population Division, "World Population Prospects, the 2015 Revision," at https://esa.un.org/unpd/wpp/.

7. Simon R. Leather, "Taxonomic Chauvinism Threatens the Future of Entomology," *Biologist* 56, no. 1 (February 2009): 10.

8. Knowing when and where a problem exists can seem boring. It is tedious. Yet increasingly it has become clear that the better we are able to know the where and when of problems, the better we will be at mitigating them. In the context of antibiotic resistance, for instance, a key challenge is knowing where bacterial pathogens resistant to antibiotics are and when they will be present. In the context of pesticide resistance, knowing where those pests are is similarly important. In all these cases, the advantage of knowing where the problem is partially comprises the ability to respond to the problem, but over the long term the advantage is more layered. It also includes the ability to plan in light of where problems are most likely to occur and to implement policies that are responsive to those companies or countries most associated with the problems. See, for example, Peter S. Jørgensen, et al., "Use Antimicrobials Wisely," *Nature* 537, no. 7619 (September 7, 2016), at http://www.nature.com/news/use-antimicrobials-wisely-1.20534.

9. In Mesopotamia, archaeologists have turned up a clay tablet from around 1800 BCE, for instance, containing advice on how best to water crops and get rid of rats. See Majda Bne Saad, "An Analysis of the Needs and Problems of Iraqi Farm Women: Implications for Agricul-

tural Extension Services" (doctoral thesis, University College, Dublin, 1990).

10. For a great short history, see Gwyn E. Jones and Chris Garforth, "The History, Development, and Future of Agricultural Extension," in *Improving Agricultural Extension: A Reference Manual*, ed. Burton E. Swanson, Robert P. Bentz, and Andrew J. Sofranko (Rome, Italy: Food and Agriculture Organization of the United Nations, 1997).

11. Erich-Christian Oerke, "Crop Losses to Pests," *The Journal of Agricultural Science* 144, no. 1 (February 2006): 31–43.

12. David Hughes, "PlantVillage: Using Smartphones and Smart Crowds for Food Security," TEDx talk (May 2015), at https://www.youtube.com/watch?v=CFWU6NoFbeA.

13. One of those scientists is Lindsay McMenemy at the Scottish Crop Research Institute, whom Hughes describes as being like "the Robin Hood of plant diseases, who robs from the rich and gives to the poor." Lindsay takes information from the publishing houses and libraries in wealthy countries and works with Hughes and Salathé to make it available to everyone.

14. Well, they give a little bit of a damn. They still spend some of their time doing lovely basic biological research on obscure ants and their fungi (Hughes) and forming beautiful evolutionary theory (Salathé). They just give much less of damn than do most.

15. Tom Ward, "Plant Village Is Reclaiming Control of Our Crops," *Huff-Post Tech*, August 18, 2013, at http://www.huffingtonpost.co.uk/tom-ward/plant-village-is-reclaiming-control-of-our-crops_b_3776047.html.

16. Nik Papageorgiou, "Smartphones to Battle Crop Disease," École Polytechnique Fédérale de Lausanne News (November 24, 2015), at http://actu.epfl.ch/news/smartphones-to-battle-crop-disease/.

17. Sharada Prasanna Mohanty, et al., "Using Deep Learning for Image-Based Plant Disease Detection," *Frontiers in Plant Science* (September 22, 2016), doi: 10.3389/fpls.2016.01419.

Epilogue

1. Julia was a postdoc with Julie Urban (then at the NC Museum of Natural Sciences) and me.

2. Statements such as "in my lab" are never really as simple as they seem. Margarita was a postdoc working with Steve Frank, David Tarpy, and me, but physically Margarita mostly worked in Dave's lab.

3. Squash bees are, simply put, wonderful. They are not social. They live solitary lives. Mother bees dig tunnels into the ground where they put

little clumps of pollen. They then lay an egg on each clump of pollen. The eggs will develop into a new generation of bees. The pollen has to be gathered though, from squash plants. During this gathering of pollen, squash bees pollinate the squash plants in the mornings and evenings when the squash flowers are open and the bees are active. Or female squash bees do anyway. Male squash bees spend their mornings (and, in some cases, evenings) lurking around squash flowers, sipping nectar, waiting for female squash bees to show up so they can mate. During the day, the male squash bees then sleep off their exertion inside the closed squash flowers. If one squeezes closed squash flowers one will often find, inside, a sleeping and slightly nectar-drunk chubby little male squash bee. Honeybees also pollinate squash flowers, but don't do so as effectively as squash bees, in part because they don't like squash flowers, in part because they are active at the wrong time of day. Margarita studied genetic differences among populations of squash bees and was able to show that the squash bees that one finds in North America moved there several thousand years ago when Native Americans took squash from Mexico and moved it north. The squash bee moved with its plant and, as a result, you can now find these bees in your backyard nearly anywhere in North America. López-Uribe, Margarita M., et al. "Crop domestication facilitated rapid geographical expansion of a specialist pollinator, the squash bee *Peponapis pruinosa*" Proc. R. Soc. B. Vol. 283. No. 1833. The Royal Society, 2016.

4. Nature is sometimes unpleasant. The beetle transmits the pathogen when it poops in the wounds it creates when it feeds. The beetle poop, and the *Erwinia* inside it, gets deep into the plant through the wound and infects its xylem. Rojas, Erika Saalau, et al. "Bacterial wilt of cucurbits: resurrecting a classic pathosystem." *Plant Disease* 99.5 (2015): 564–574.

5. Bottle gourds were domesticated in Africa and spread around the world in no small part because people liked to use the gourds as water containers (the bottle gourd was, somehow, already in the Americas, for instance, when Columbus arrived).

6. Morimoto, Yasuyuki, Mary Gikungu, and Patrick Maundu. "Pollinators of the bottle gourd (*Lagenaria siceraria*) observed in Kenya." *International Journal of Tropical Insect Science* 24.01 (2004): 79–86.

Index

About the Author

Rob Dunn is a professor in the department of applied ecology at North Carolina State University and in the Center for Macroecology, Evolution, and Climate at the Danish Natural History Museum. He is the author of *The Man Who Touched His Own Heart, The Wild Life of Our Bodies,* and *Every Living Thing,* and his magazine work is published widely, including in *National Geographic, Natural History, New Scientist, Scientific American,* and *Smithsonian.* He has a PhD from the University of Connecticut and was a Fulbright fellow. He lives in Raleigh, North Carolina.